SpringerBriefs in Optimization

SpringerBriefs in Optimization showcases algorithmic and theoretical techniques, case studies, and applications within the broad-based field of optimization. Manuscripts related to the ever-growing applications of optimization in applied mathematics, engineering, medicine, economics, and other applied sciences are encouraged.

For further volumes:
http://www.springer.com/series/8918

Khanh D. Pham

Linear-Quadratic Controls in Risk-Averse Decision Making

Performance-Measure Statistics and Control Decision Optimization

Springer

Khanh D. Pham
The Air Force Research Laboratory
Space Vehicles Directorate
3550 Aberdeen Ave. S.E.
Kirtland Air Force Base
New Mexico, USA

ISSN 2190-8354 ISSN 2191-575X (electronic)
ISBN 978-1-4614-5078-8 ISBN 978-1-4614-5079-5 (eBook)
DOI 10.1007/978-1-4614-5079-5
Springer New York Heidelberg Dordrecht London

Library of Congress Control Number: 2012949189

Mathematics Subject Classification (2010): 60-00, 93cxx, 62-xx, 65k10

Printed on acid-free paper

Springer is part of Springer Science+Business Media (www.springer.com)

Dedicated to my Wife Huong Nguyen
and my Children An and Duc

Preface

When measuring performance reliability, statistical analysis for probabilistic nature of performance uncertainty is relied on as part of the long-range assessment of reliability. Some of the most widely used measures for performance reliability are mean and variance which attempt to summarize the underlying performance variations. However, other aspects of performance distributions, which do not appear in most of the existing progress, are skewness, flatness, etc. Recent developments connecting statistics with decision optimization have shown that some performance with negative skewness appears riskier than performance with positive skewness, when expectation and variance are held constant. If skewness does, indeed, play an essential role in determining the perception of risks, then the range of applicability of the present theory for stochastic control and decision analysis should be restricted, e.g., to symmetric or equally skewed performance measures.

Using these findings as a point of departure, one of the goals of writing this monograph is exactly to take risk preferences of the feedback control designer and/or the decision maker into account. Unlike other books that cover the state of the art in stochastic control theory, this monograph cuts across control engineering (control feedback and decision optimization) and statistics (post-design performance analysis) with a common theme: reliability increase seen from the responsive angle of incorporating and engineering multilevel performance robustness beyond the long-run average performance into control feedback design and decision making and complex dynamic systems from the start. The basic approach is to effectively integrate performance-information analysis into risk-averse feedback control selection for performance robustness and reliability requirements so that (1) intrinsic performance variations caused by process noise stochasticity from exogenous disturbances and uncertain environments are addressed concurrently with other high performance demands and (2) trade-off analysis on performance risks and values directly evaluates the impact of reliability as well as other performance requirements. Hence, via higher-order performance-measure statistics and adaptive decision making, it is anticipated that future performance variations will lose the element of surprise due to the inherent property of self-enforcing and risk-averse decision solutions, which are highly capable of reshaping the cumulative probability distribution of closed-loop performance.

Given the aforementioned background, this monograph provides a complete description of statistical optimal control (also known as cost-cumulant control) theory. In this viewpoint, this theory unifies decisions with risk consequences and feedback control designs for a class of stochastic linear dynamical systems with performance appraisals of finite-horizon integral-quadratic-form random costs. In writing up different control problems and topics, emphasis is primarily placed on major developments attained, thus and then explicit connections between mathematical statistics of performance appraisals and decision and control optimization with performance risk aversion are made for: the stochastic linear system dynamics subjected to either general, closely supervised, servo, or model-following tracking criteria; the focus of perfect state-feedback measurements and corrupted output-feedback observations; exemplification of closed-loop system trajectory sensitivity with respect to constant parameters as well. Each chapter is supplemented by an extended bibliography and indexes.

Throughout the monograph, the methodology taken herein is composed of five steps: (1) problem descriptions in which basic assumptions related to the state-space models are discussed; (2) mathematical statistics for performance robustness whose backward differential equations are characterized by making use of both compactness from logics of state-space models and quantitativity from a-priori knowledge of probabilistic processes, (3) statements of statistical optimal control which provide complete problem formulations composed of unique notations, terminologies, definitions, and theorems; (4) existence of control solutions whereby information about accuracy of estimates for higher-order characteristics of performance-measure uncertainty is incorporated into risk-averse decision strategies; and (5) chapter summaries which are given to shed some lights on the relevance of the developed results.

It is planned while organizing the materials that this monograph would be appropriate for use either as graduate-level lectures in applied mathematics and electrical engineering with systems-theoretic concentration, a good volume for elective study or a reference for interested readers, researchers, and graduate students who are interested in theoretical constructs and design principles for stochastic controlled systems to include the requirements of performance reliability and decision making with risk consequences and emerging effects within certain stochastic environments to determine the understanding of performance variations, risk-averse attitudes, and the course correction required for realistic situations.

Finally, the present work is originated from the early research, i.e., the cost-cumulant control which owes its existence, primarily, to the influence of Prof. Stanley R. Liberty and Prof. Michael K. Sain with whom the author had the pleasure of collaboration during the graduate studies at University of Nebraska and University of Notre Dame. Indeed, this monograph is an outgrowth of my academic research activities over the past few years after joining the Air Force Research Laboratory-Space Vehicles Directorate. In this regard, the highly stimulated research environment at the Spacecraft Components Technology branch is gratefully acknowledged. Next I am deeply appreciative of the Air Force Office of Scientific Research for funding support in parts of this work through laboratory research initiative requests. Last but not least, no amount of thanks would suffice for the cheerful and encouraging support I have received from my wife, Huong, and children, An and Duc, on the domestic front where I am allowed to use my spare time for writing this monograph.

Albuquerque, NM, USA — Khanh D. Pham

Contents

1 Introduction 1
1.1 Risk Assessment and Management 1
1.2 Monograph Ideas and Contributions 2
1.3 Methodology 2
1.4 Chapter Organization 3
References 5

2 Risk-Averse Control of Linear-Quadratic Tracking Problems 7
2.1 Introduction 7
2.2 The Problem 8
2.3 The Value of Performance-Measure Statistics 9
2.4 Statements of Mayer Problem with Performance Risk Aversion 16
2.5 Risk-Averse Control of Adaptive Behavior 21
2.6 Chapter Summary 28
References 29

3 Overtaking Tracking Problems in Risk-Averse Control 31
3.1 Introduction 31
3.2 Problem Description 32
3.3 A Framework for Performance-Measure Statistics 34
3.4 Statements of the Risk-Averse Control Problem 38
3.5 Optimal Risk-Averse Tracking Solution 42
3.6 Chapter Summary 46
References 47

4 Performance Risk Management in Servo Systems 49
4.1 Introduction 49
4.2 The Performance Information Process 50
4.3 The System Control Problem 54
4.4 Statistical Optimal Control Solution 59
4.5 Chapter Summary 65
References 66

5 Risk-Averse Control Problems in Model-Following Systems 67
5.1 Introduction 67
5.2 Performance Information in Control with Risk Consequences 68
5.3 Formulation of the Control Problem 75
5.4 Existence of Risk-Averse Control Solution 80
5.5 Chapter Summary 85
References 85

6 Incomplete Feedback Design in Model-Following Systems 87
6.1 Introduction 87
6.2 Backward Differential Equations for Performance-Measure Statistics 88
6.3 Performance-Measure Statistics for Risk-Averse Control 95
6.4 Output-Feedback Control Solution 100
6.5 Chapter Summary 104
References 104

7 Reliable Control for Stochastic Systems with Low Sensitivity 107
7.1 Introduction 107
7.2 The Problem and Representations for Performance Robustness...... 108
7.3 Problem Statements with the Maximum Principle 114
7.4 Low Sensitivity Control with Risk Aversion 118
7.5 Chapter Summary 122
References 123

8 Output-Feedback Control for Stochastic Systems with Low Sensitivity 125
8.1 Introduction 125
8.2 Linking Performance-Measure Statistics and Risk Perceptions 126
8.3 Statements Toward the Optimal Decision Problem.................... 136
8.4 Risk-Averse Control Solution in the Closed-Loop System 141
8.5 Chapter Summary 144
References 144

9 Epilogue 145
9.1 The Aspect of Performance-Information Analysis 145
9.2 The Aspect of Risk-Averse Decision Making 146
9.3 What Will Be?.. 146

Index 149

Chapter 1
Introduction

Abstract In introducing the idea of early risk planning, Sain [1] suggested that understanding the dynamics of performance cost in a stochastic control problem requires the establishment of risk-monitoring metrics. These metrics, including but not limited to the cumulants of such cost functions, allow measurement and evaluation of the status of risk-handling options and system performance. In addition, Stanley R. Liberty was the first person to ever calculate and display performance-measure probability densities as published with Michael K. Sain in the research article titled, "Performance-Measure Densities for a Class of linear-quadratic Gaussian (LQG) Control Systems," which appeared in Sain and Liberty [2]. These ideas became the originating node for the author's research investigation on linear-quadratic controls in risk-averse decision making, a theme which will be evident enough in this monograph.

1.1 Risk Assessment and Management

For the purposes of this introduction, it is sufficient to note that the influence of this monograph is present both in risk assessment and risk management pertaining to system analysis and control synthesis, whereby optimal control of a random process naturally raises the question of the cost functional as a random variable. The former perspective emphasizes the dual role of identification and transformation of performance uncertainty into quantified risk such as cost cumulants that are based upon the series expansion of the logarithm of the cost characteristic function. The latter perspective focuses on the aspect of risk handling and concentrates on the feedback control mechanism of shaping of a probability density function of the random cost variable that is therefore responsible for increasing performance robustness and reliability. Elsewhere on the existing publications, it is possible to sample the way that the subject research by the author looks in real-world applications, such as seismic protection of buildings and bridges from earthquakes,

K.D. Pham, *Linear-Quadratic Controls in Risk-Averse Decision Making*,
SpringerBriefs in Optimization, DOI 10.1007/978-1-4614-5079-5_1,

winds, and seas, by means of the IEEE and ASME conference proceedings, Journal of Optimization Theory and Applications, and chapters in Birkhauser and Springer books.

1.2 Monograph Ideas and Contributions

During the past several decades, there have been real-world applications for which the associated control design problems and post-design performance analyses cannot be solved by using one-shot approaches. In this regard, the main value of this monograph rests in the diversity of efforts to derive some of the consequences of what makes performance-measure statistics meaningful and relevant risk-averse feedback control strategies of analysis: the interdependence of performance uncertainty quantification and feedback controls and decisions with risk-averse attitudes. Over the years, such approaches have been dynamically evolved into a body of "statistical optimal control theories". This monograph is written about the wide spectrum of statistical optimal control theories. In brief, the main contributions of the monograph are the following: (1) it provides a general framework to identify and transform performance-measure uncertainty associated with various classes of linear-quadratic stochastic systems into mathematical statistics for post-design analysis; (2) it addresses several issues like the benefits of feedback and performance-measure statistics in multi-input/multi-output dynamical systems, risk-value aware performance criteria, and risk-averse decision making; (3) it presents the key concept of performance-measure statistics with its proofs followed by efficient computational methods; and (4) it establishes risk-averse feedback control techniques.

1.3 Methodology

Throughout the monograph, the methodology in each chapter/section is composed of five steps:

- Mathematical modeling in which basic assumptions related to the state-space models are considered.
- Mathematical statistics for performance robustness whose backward differential equations are characterized by making use of both compactness from logics of state-space models and quantitativity from a priori knowledge of probabilistic processes.
- Statements of statistical optimal control which provide complete problem formulations composed of unique notations, terminologies, definitions, and theorems to pave the way for subsequent analysis.

- Existence of control solutions whereby information about accuracy of estimates for higher-order characteristics of performance-measure uncertainty is incorporated into risk-averse decision strategies. The results are provided most of the time in the form of theorems, lemmas, and corollaries.
- Chapter Summaries which are given to shed some light of the relevance of the developed results.

For convenience, the relevant references are offered at the end of each chapter for the purpose of stimulating the reader. It is hoped that this way of articulating the information will attract the attention of a wide spectrum of readership.

1.4 Chapter Organization

As will be seen in the monograph, the shape and functional form of a utility function tell a great deal about the basic attitudes of control designers toward the uncertain outcomes or performance risks. Of particular, the new utility function or the so-called the risk-value aware performance index, which is being proposed herein as a linear manifold defined by a finite number of centralized moments associated with a random quadratic cost functional will provide a convenient allocation representation of apportioning performance robustness and reliability requirements into the multi-attribute requirement of qualitative characteristics of expected performance and performance risks. The literature grew progressively, and quite a number of fundamental concepts and powerful tools have been developed when a measure of uncertainty such as the average and variance values of the subject performance measure was used. Despite the existing progress so far, many investigations have indicated that any evaluation scheme based on just the first two statistics of performance measure would necessarily imply indifference between some courses of action; therefore, no criterion based solely on the first two attributes of means and variance correctly represents their preferences. This long-standing technical challenge provided the motivation to write this current monograph. This monograph presents theoretical explorations on several fundamental problems for stochastic linear-quadratic systems.

This monograph is primarily intended for researchers and engineers in the systems and control community. It can also serve as complementary reading for linear system theory at the postgraduate level. After the first introductory chapter, the core material of the monograph is organized as follows.

Chapter 2 tackles an optimal control problem for a class of tracking systems. The control objective is to minimize a finite, linear combination of cumulants of integral quadratic cost over linear, memoryless, full-state-feedback control laws. Perhaps the implication of the feasibility of the solution to the statistical optimal control is supported by a set of coupled Riccati-type differential equations. And the conditions on the existence of solutions are also derived.

Borrowing from the research on linear-quadratic tracking in Chap. 2, the central problem of this view in Chap. 3 is interpreted as the problem of simultaneously track reference trajectory and command input in accordance of the so-called risk-averse performance index. In the synthesis of statistical optimal controllers, the possibility of translating risk-averse attitudes into performance reliability typically involves some combination of accurate prediction of the effects of chi-squared randomness on performance distribution of the optimal tracking and the corresponding feedback control synthesis.

The problem addressed in Chap. 4 is the linear-quadratic class of servo systems. The notion of risk-averse feedback control will be defined and discussed in detail later; however, for the time being, a control is said to be risk aversive if it continues to provide performance reliability against all the sample realizations of the stochastic process noises and regulation under perturbation of reference model and the system outputs.

In the most general terms, Chaps. 5 and 6 identify and explore theoretical narratives of risk-averse control for a linear-quadratic class of model-following control systems. The basic idea is based on the fact that performance information with higher-order performance-measure statistics can improve control decisions for closed-loop system performance reliability, but the controller design can also be computationally involved. Starting from the similar considerations, the theory obtained herein is further extended to the case of model-following control systems with incomplete state feedback in Chap. 6. It is shown that performance information can improve control decisions with only available output measurements for system performance reliability.

As a part of a more general tendency in the study of statistical optimal control theory, Chap. 7 investigates the problem of controlling stochastic linear systems with quadratic criterion which includes sensitivity variables. An optimal risk-averse control strategy with low sensitivity has been achieved by the cascade of mathematical statistics of performance uncertainty and a linear feedback. Furthermore, necessary and sufficient conditions for the optimal solution are given as set of nonlinear matrix equations. Despite obvious differences in the unit of analysis, the results are extended to the case when only noisy measurements are available in Chap. 8, where an optimal risk-averse control strategy with low sensitivity and subject to noisy feedback measurements is investigated. It is shown that the certainty equivalence principle still holds for this statistical optimal control problem.

Finally, Chap. 9 contains a precise summary with which one might reflect two striking perspectives; e.g., (1) the acquisition and utilization of insights, regarding whether the control designers are risk averse or risk prone and thus restrictions on the utility functions implied by these attitudes, to adapt feedback control strategies to meet austere environments and (2) best courses of action to ensure performance robustness and reliability, provided that the control designers be subscribed to certain attitudes that all the chapters in this monograph share and contribute defining.

References

1. Sain, M.K.: Control of linear systems according to the minimal variance criterion-a new approach to the disturbance problem. IEEE Trans. Automat. Contr. **11**, 118–122 (1966)
2. Sain, M.K., Liberty, S.R.: Performance measure densities for a class of LQG control systems. IEEE Trans. Automat. Contr. **16**, 431–439 (1971)

Chapter 2
Risk-Averse Control of Linear-Quadratic Tracking Problems

Abstract The topic of risk-averse control is currently receiving substantial research from the theoretical community oriented toward stochastic control theory. For instance, this present chapter extends the application of risk-averse controller design to control a wide class of linear-quadratic tracking systems where output measurements of a tracker follow as closely as possible a desired trajectory via a complete statistical description of the associated integral-quadratic performance measure. It is shown that the tracking problem can be solved in two parts: one, a feedback control whose optimization criterion is based on a linear combination of finite cumulant indices of an integral-quadratic performance measure associated to a linear stochastic tracking system over a finite horizon and two, an affine control which takes into account of dynamics mismatched between a desired trajectory and tracker states.

2.1 Introduction

An interesting extension of the statistical optimal control [1–16] when both perfect and noisy state measurements are available is to make a linear stochastic system track as closely as possible a desired trajectory via a complete statistical description of the associated finite-horizon integral-quadratic-form cost. To the best knowledge of the author, this theoretical development appears to be the first of its kind, and the optimal control problem being considered herein is actually quite general and will enable control engineer not only to penalize for variations in, as well as for the levels of, the state variables and control variables but also to characterize the probabilistic distribution of the performance measure as needed in post controller-design analysis. Since this problem formulation is parameterized both by the number of cumulants and by the scalar coefficients in the linear combination, it defines a very general linear-quadratic-Gaussian (LQG) and risk sensitive problem classes. The special cases where only the first cost cumulant is minimized and whereas a denumerable linear combination of cost cumulants is minimized are, of course, the

K.D. Pham, *Linear-Quadratic Controls in Risk-Averse Decision Making*,
SpringerBriefs in Optimization, DOI 10.1007/978-1-4614-5079-5_2,

well-known minimum-mean LQG problem and the risk sensitive control objective, respectively. Some practical applications for this theoretical development can be found in [17] where in tactical and combat situations, a vehicle with the goal-seeking nature initially decides on an appropriate destination and then moves in an optimal fashion toward that destination.

In the rest of this chapter, the problem and the value of performance-measure statistics of the linear-quadratic class of tracking problems are presented in Sects. 2.2 and 2.3. The statements of deterministic optimization of Mayer type and performance risk aversion are discussed in Sect. 2.4. The risk-averse control solution of adaptive behavior is evident in Sect. 2.5, while some conclusions are drawn in Sect. 2.6.

2.2 The Problem

Consider a linear stochastic tracking system governed by the following controlled diffusion process:

$$dx(t) = (A(t)x(t) + B(t)u(t))dt + G(t)dw(t), \quad x(t_0) = x_0, \tag{2.1}$$

$$y(t) = C(t)x(t), \tag{2.2}$$

where the continuous-time coefficients $A \in \mathscr{C}([t_0,t_f];\mathbb{R}^{n\times n})$, $B \in \mathscr{C}([t_0,t_f];\mathbb{R}^{n\times m})$, $C \in \mathscr{C}([t_0,t_f];\mathbb{R}^{r\times n})$, and $G \in \mathscr{C}([t_0,t_f];\mathbb{R}^{n\times p})$. The system noise $w(t) \in \mathbb{R}^p$ is the p-dimensional stationary Wiener process starting from t_0, independent of the known initial-value condition x_0, and defined on a complete probability space $(\Omega, \mathscr{F}, \mathscr{P})$ over $[t_0,t_f]$ with the correlation of independent increments

$$E\left\{[w(\tau_1) - w(\tau_2)][w(\tau_1) - w(\tau_2)]^T\right\} = W|\tau_1 - \tau_2|, \quad W > 0, \quad \forall \tau_1, \tau_2 \in [t_0,t_f].$$

The feasible control input $u \in L^2_{\mathscr{F}_t}(\Omega; \mathscr{C}([t_0,t_f];\mathbb{R}^m))$ the subset of Hilbert space of $\mathbb{R}^m$-valued square-integrable process on $[t_0,t_f]$ that are adapted to the σ-field $\mathscr{F}_t$ generated by $w(t)$ to the specified system model is selected so that the resulting output $y \in L^2_{\mathscr{F}_t}(\Omega; \mathscr{C}([t_0,t_f];\mathbb{R}^r))$ best matches the desired output $\zeta \in L^2([t_0,t_f];\mathbb{R}^r)$.

Associated with admissible initial condition (t_0, x_0) and feasible control input u is a traditional integral-quadratic form cost $J : L^2_{\mathscr{F}_t}(\Omega; \mathscr{C}([t_0,t_f];\mathbb{R}^m)) \mapsto \mathbb{R}^+$

$$J(u(\cdot)) = (y(t_f) - \zeta(t_f))^T Q_f (y(t_f) - \zeta(t_f)) + \int_{t_0}^{t_f} [(y(\tau) - \zeta(\tau))^T Q(\tau)(y(\tau) - \zeta(\tau)) + u^T(\tau)R(\tau)u(\tau)]d\tau \tag{2.3}$$

in which the terminal penalty error weighting $Q_f \in \mathbb{R}^{r\times r}$, the error weighting $Q \in \mathscr{C}([t_0,t_f];\mathbb{R}^{r\times r})$, and the control input weighting $R \in \mathscr{C}([t_0,t_f];\mathbb{R}^{m\times m})$ are symmetric and positive semidefinite with $R(t)$ invertible.

In the perfect-state measurement case, the initial system state is assumed to be known exactly, and the control input is generated by a closed-loop control policy of interest $\gamma : [t_0,t_f] \times L^2_{\mathscr{F}_t}(\Omega;\mathscr{C}([t_0,t_f];\mathbb{R}^n)) \mapsto L^2_{\mathscr{F}_t}(\Omega;\mathscr{C}([t_0,t_f];\mathbb{R}^m))$, according to the control law

$$u(t) = \gamma(t,x(t)) \triangleq K(t)x(t) + l(t), \tag{2.4}$$

where $l \in \mathscr{C}([t_0,t_f];\mathbb{R}^m)$ is an affine control signal and $K \in \mathscr{C}([t_0,t_f];\mathbb{R}^{m\times n})$ is an admissible feedback gain in a sense to be specified in later sections as appropriate.

Henceforth, for the given initial condition $(t_0,x_0) \in [t_0,t_f] \times \mathbb{R}^n$ and subject to the control policy (2.4), the dynamics of the tracking problem are governed by

$$dx(t) = [A(t) + B(t)K(t)]x(t)dt + B(t)l(t)dt + G(t)dw(t), \quad x(t_0) = x_0, \tag{2.5}$$

$$y(t) = C(t)x(t) \tag{2.6}$$

and the finite-horizon integral-quadratic-form cost

$$J(K,l) = (y(t_f) - \zeta(t_f))^T Q_f (y(t_f) - \zeta(t_f)) + \int_{t_0}^{t_f} [(y(\tau) - \zeta(\tau))^T Q(\tau)(y(\tau) - \zeta(\tau)) + (K(\tau)x(\tau) + l(\tau))^T R(\tau)(K(\tau)x(\tau) + l(\tau))]d\tau. \tag{2.7}$$

2.3 The Value of Performance-Measure Statistics

A rather innovative approach of coping with the uncertain cost (2.7) of the chi-squared type is to use the mathematical statistics of the random cost (2.7) in a responsive risk-value aware performance index as appropriately required in statistical optimal control herein and convert the output tracking problem subject to control decisions with risk consequences into a deterministic controller design problem and then solve it. The following development characterizes how the mathematical statistics of the random cost (2.7) are formulated. In particular, the initial condition (t_0,x_0) is replaced by any arbitrary pair (τ,x_τ). Then, for the given affine input l and admissible feedback gain K, the cost functional (2.7) is seen as the "cost-to-go," $J(\tau,x_\tau)$. The moment-generating function of the vector-valued random process (2.5) is given by the definition

$$\varphi(\tau,x_\tau;\theta) \triangleq E\{\exp(\theta J(\tau,x_\tau))\}, \tag{2.8}$$

where the scalar $\theta \in \mathbb{R}^+$ is a small parameter. Thus, the cumulant-generating function immediately follows

$$\psi(\tau,x_\tau;\theta) \triangleq \ln\{\varphi(\tau,x_\tau;\theta)\} \tag{2.9}$$

in which $\ln\{\cdot\}$ denotes the natural logarithmic transformation of an enclosed entity.

Theorem 2.3.1. *Let* $\varphi(\tau,x_\tau;\theta) \triangleq \rho(\tau,\theta)\exp\{x_\tau^T \Upsilon(\tau,\theta)x_\tau + 2x_\tau^T\eta(\tau,\theta)\}$ *and further let* $\upsilon(\tau,\theta) \triangleq \ln\{\rho(\tau,\theta)\}$*, for all* $\tau \in [t_0,t_f]$ *and* $\theta \in \mathbb{R}^+$*. Then, the cumulant-generating function is given by*

$$\psi(\tau,x_\tau;\theta) = x_\tau^T \Upsilon(\tau,\theta)x_\tau + 2x_\tau^T\eta(\tau,\theta) + \upsilon(\tau,\theta) \tag{2.10}$$

whereby the cumulant building variables $\Upsilon(\tau,\theta)$*,* $\eta(\tau,\theta)$ *and* $\upsilon(\tau,\theta)$ *solve the backward-in-time differential equations*

$$\begin{aligned}\frac{d}{d\tau}\Upsilon(\tau,\theta) = &-[A(\tau)+B(\tau)K(\tau)]^T\Upsilon(\tau,\theta) - \Upsilon(\tau,\theta)[A(\tau)+B(\tau)K(\tau)]\\ &-2\Upsilon(\tau,\theta)G(\tau)WG^T(\tau)\Upsilon(\tau,\theta)\\ &-\theta(C^T(\tau)Q(\tau)C(\tau)+K^T(\tau)R(\tau)K(\tau)),\end{aligned} \tag{2.11}$$

$$\begin{aligned}\frac{d}{d\tau}\eta(\tau,\theta) = &-[A(\tau)+B(\tau)K(\tau)]^T\eta(\tau,\theta) - \Upsilon(\tau,\theta)B(\tau)l(\tau)\\ &-\theta K^T(\tau)R(\tau)l(\tau) + \theta C^T(\tau)Q(\tau)\zeta(\tau),\end{aligned} \tag{2.12}$$

$$\begin{aligned}\frac{d}{d\tau}\upsilon(\tau,\theta) = &-\mathrm{Tr}\{\Upsilon(\tau,\theta)G(\tau)WG^T(\tau)\}\\ &-2\eta^T(\tau,\theta)B(\tau)l(\tau) - \theta l^T(\tau)R(\tau)l(\tau) - \theta\zeta^T(\tau)Q(\tau)\zeta(\tau)\end{aligned} \tag{2.13}$$

together with the terminal-value conditions $\Upsilon(t_f,\theta) = \theta C^T(t_f)Q_fC(t_f)$*,* $\eta(t_f,\theta) = \theta C^T(t_f)Q_f\zeta(t_f)$ *and* $\upsilon(t_f,\theta) = \theta\zeta^T(t_f)Q_f\zeta(t_f)$*.*

Proof. For any $\theta \in \mathbb{R}^+$ given and $\varpi(\tau,x_\tau;\theta) \triangleq \exp\{\theta J(\tau,x_\tau)\}$, the moment-generating function therefore becomes $\varphi(\tau,x_\tau;\theta) = E\{\varpi(\tau,x_\tau;\theta)\}$ with the time derivative of

$$\begin{aligned}\frac{d}{d\tau}\varphi(\tau,x_\tau;\theta) = -\theta\Big\{&x_\tau^T[C^T(\tau)Q(\tau)C(\tau)+K^T(\tau)R(\tau)K(\tau)]x_\tau\\ &+2x_\tau^T[K^T(\tau)R(\tau)l(\tau) - C^T(\tau)Q(\tau)\zeta(\tau)]\\ &+l^T(\tau)R(\tau)l(\tau) + \zeta^T(\tau)Q(\tau)\zeta(\tau)\Big\}\varphi(\tau,x_\tau;\theta).\end{aligned} \tag{2.14}$$

Using the standard Ito's formula, it yields

$$\begin{aligned}d\varphi(\tau,x_\tau;\theta) =& E\{d\varpi(\tau,x_\tau;\theta)\}\\ =& E\Big\{\varpi_\tau(\tau,x_\tau;\theta)\,d\tau + \varpi_{x_\tau}(\tau,x_\tau;\theta)\,dx_\tau\\ &+\frac{1}{2}\mathrm{Tr}\{\varpi_{x_\tau x_\tau}(\tau,x_\tau;\theta)G(\tau)WG^T(\tau)\}\,d\tau\Big\}\\ =& \varphi_\tau(\tau,x_\tau;\theta)\,d\tau + \varphi_{x_\tau}(\tau,x_\tau;\theta)\,[(A(\tau)+B(\tau)K(\tau))x_\tau + B(\tau)l(\tau)]d\tau\\ &+\frac{1}{2}\mathrm{Tr}\{\varphi_{x_\tau x_\tau}(\tau,x_\tau;\theta)\,G(\tau)WG^T(\tau)\}\,d\tau\end{aligned}$$

from which, by the definition $\varphi(\tau, x_\tau; \theta) \triangleq \rho(\tau, \theta) \exp\{x_\tau^T \Upsilon(\tau, \theta) x_\tau + 2x_\tau^T \eta(\tau, \theta)\}$ and its partial derivatives

$$\varphi_\tau(\tau, x_\tau; \theta) = \left[\frac{\frac{d}{d\tau}\rho(\tau, \theta)}{\rho(\tau, \theta)} + x_\tau^T \frac{d}{d\tau}\Upsilon(\tau, \theta) x_\tau + 2x_\tau^T \frac{d}{d\tau}\eta(\tau, \theta) \right] \varphi(\tau, x_\tau; \theta),$$

$$\varphi_{x_\tau}(\tau, x_\tau; \theta) = \left\{ x_\tau^T \left[\Upsilon(\tau, \theta) + \Upsilon^T(\tau, \theta) \right] + 2\eta^T(\tau, \theta) \right\} \varphi(\tau, x_\tau; \theta),$$

$$\varphi_{x_\tau x_\tau}(\tau, x_\tau; \theta) = \left[\Upsilon(\tau, \theta) + \Upsilon^T(\tau, \theta) \right] \varphi(\tau, x_\tau; \theta) \\ + \left[\Upsilon(\tau, \theta) + \Upsilon^T(\tau, \theta) \right] x_\tau x_\tau^T \left[\Upsilon(\tau, \theta) + \Upsilon^T(\tau, \theta) \right] \varphi(\tau, x_\tau; \theta)$$

leads to

$$\frac{d}{d\tau}\varphi(\tau, x_\tau; \theta) \\ = \left\{ \frac{\frac{d}{d\tau}\rho(\tau, \theta)}{\rho(\tau, \theta)} + 2x_\tau^T \Upsilon(\tau, \theta) G(\tau) W G^T(\tau) \Upsilon(\tau, \theta) x_\tau \right. \\ + x_\tau^T \left[\frac{d}{d\tau}\Upsilon(\tau, \theta) + (A(\tau) + B(\tau)K(\tau))^T \Upsilon(\tau, \theta) + \Upsilon(\tau, \theta)(A(\tau) + B(\tau)K(\tau)) \right] x_\tau \\ + 2x_\tau^T \left[\frac{d}{d\tau}\eta(\tau, \theta) + (A(\tau) + B(\tau)K(\tau))^T \eta(\tau, \theta) + \Upsilon(\tau, \theta) B(\tau) l(\tau) \right] \\ \left. + \operatorname{Tr}\left\{ \Upsilon(\tau, \theta) G(\tau) W G^T(\tau) \right\} + 2\eta^T(\tau, \theta) B(\tau) l(\tau) \right\} \varphi(\tau, x_\tau; \theta). \quad (2.15)$$

By applying the result (2.14) to this finding (2.15) and having both linear and quadratic terms independent of the arbitrary x_τ, it requires that

$$\frac{d}{d\tau}\Upsilon(\tau, \theta) = -\left[A(\tau) + B(\tau)K(\tau)\right]^T \Upsilon(\tau, \theta) - \Upsilon(\tau, \theta)\left[A(\tau) + B(\tau)K(\tau)\right] \\ - 2\Upsilon(\tau, \theta) G(\tau) W G^T(\tau) \Upsilon(\tau, \theta) \\ - \theta C^T(\tau) Q(\tau) C(\tau) - \theta K^T(\tau) R(\tau) K(\tau),$$

$$\frac{d}{d\tau}\eta(\tau, \theta) = -\left[A(\tau) + B(\tau)K(\tau)\right]^T \eta(\tau, \theta) - \Upsilon(\tau, \theta) B(\tau) l(\tau) \\ - \theta K^T(\tau) R(\tau) l(\tau) + \theta C^T(\tau) Q(\tau) \zeta(\tau),$$

$$\frac{d}{d\tau}\upsilon(\tau, \theta) = -\operatorname{Tr}\left\{ \Upsilon(\tau, \theta) G(\tau) W G^T(\tau) \right\} - 2\eta^T(\tau, \theta) B(\tau) l(\tau) \\ - \theta l^T(\tau) R(\tau) l(\tau) - \theta \zeta^T(\tau) Q(\tau) \zeta(\tau).$$

At the final time $\tau = t_f$, it follows that

$$\varphi(t_f, x(t_f); \theta) = \rho(t_f, \theta) \exp\{x^T(t_f) \Upsilon(t_f, \theta) x(t_f) + 2x^T(t_f) \eta(t_f, \theta)\}$$
$$= E\{\exp\{\theta [y(t_f) - \zeta(t_f)]^T Q_f [y(t_f) - \zeta(t_f)]\}\}$$

which in turn yields the terminal-value conditions: $\Upsilon(t_f, \theta) = \theta C^T(t_f) Q_f C(t_f)$, $\eta(t_f, \theta) = -\theta C^T(t_f) Q_f \zeta(t_f)$, $\rho(t_f, \theta) = \exp\{\theta \zeta^T(t_f) Q_f \zeta(t_f)\}$ and $\upsilon(t_f, \theta) = \theta \zeta^T(t_f) Q_f \zeta(t_f)$. □

By definition, the mathematical statistics associated with the random cost (2.7) are now generated by the use of a Maclaurin series expansion for the cumulant-generating function

$$\psi(\tau, x_\tau; \theta) \triangleq \sum_{r=1}^{\infty} \kappa_r \frac{\theta^r}{r!} = \sum_{r=1}^{\infty} \frac{\partial^r}{\partial \theta^r} \psi(\tau, x_\tau; \theta)\bigg|_{\theta=0} \frac{\theta^r}{r!} \tag{2.16}$$

in which κ_r are called the rth performance-measure statistics or cost cumulants.

In addition, the coefficients of the Maclaurin series expansion are computed by using the result (2.10)

$$\frac{\partial^r}{\partial \theta^r} \psi(\tau, x_\tau; \theta)\bigg|_{\theta=0} = x_\tau^T \frac{\partial^r}{\partial \theta^r} \Upsilon(\tau, \theta)\bigg|_{\theta=0} x_\tau$$
$$+ 2x_\tau^T \frac{\partial^r}{\partial \theta^r} \eta(\tau, \theta)\bigg|_{\theta=0} + \frac{\partial^r}{\partial \theta^r} \upsilon(\tau, \theta)\bigg|_{\theta=0}. \tag{2.17}$$

In view of the results (2.16) and (2.17), the cost cumulants for the linear-quadratic tracking problem with performance uncertainty are then described as follows:

$$\kappa_r = x_\tau^T \frac{\partial^r}{\partial \theta^r} \Upsilon(\tau, \theta)\bigg|_{\theta=0} x_\tau + 2x_\tau^T \frac{\partial^r}{\partial \theta^r} \eta(\tau, \theta)\bigg|_{\theta=0} + \frac{\partial^r}{\partial \theta^r} \upsilon(\tau, \theta)\bigg|_{\theta=0} \tag{2.18}$$

for any finite $1 \le r < \infty$.

For notational convenience, the definition that follows is essential, e.g., $H(\tau, r) \triangleq \frac{\partial^r}{\partial \theta^r} \Upsilon(\tau, \theta)\Big|_{\theta=0}$, $\breve{D}(\tau, i) \triangleq \frac{\partial^r}{\partial \theta^r} \eta(\tau, \theta)\Big|_{\theta=0}$ and $D(\tau, r) \triangleq \frac{\partial^r}{\partial \theta^r} \upsilon(\tau, \theta)\Big|_{\theta=0}$. Then, the performance-measure statistics associated with the linear-quadratic tracking problem considered herein exhibits a familiar pattern of quadratic affine in arbitrary initial system state x_τ as will be illustrated in more detail in the next result.

Theorem 2.3.2 (Performance-Measure Statistics). *The linear system dynamics governed by Eqs. (2.5)–(2.6) attempt to track the prescribed signal $\zeta(t)$ with the chi-squared random cost (2.7). The kth performance-measure statistic considered for performance risk aversion is given by*

$$\kappa_k = x_0^T H(t_0, k) x_0 + 2x_0^T \breve{D}(t_0, k) + D(t_0, k), \quad k \in \mathbb{Z}^+ \tag{2.19}$$

in which the building variables $\{H(\tau,r)\}_{r=1}^{k}$, $\{\breve{D}(\tau,r)\}_{r=1}^{k}$ *and* $\{D(\tau,r)\}_{r=1}^{k}$ *evaluated at* $\tau = t_0$ *satisfy the time-backward differential equations (with the dependence of* $H(\tau,r)$, $\breve{D}(\tau,r)$ *and* $D(\tau,r)$ *upon the addmisbile* l *and* K *suppressed)*

$$\frac{d}{d\tau}H(\tau,1) = -\left[A(\tau)+B(\tau)K(\tau)\right]^T H(\tau,1) - H(\tau,1)\left[A(\tau)+B(\tau)K(\tau)\right]$$
$$- C^T(\tau)Q(\tau)C(\tau) - K^T(\tau)R(\tau)K(\tau), \tag{2.20}$$

$$\frac{d}{d\tau}H(\tau,r) = -\left[A(\tau)+B(\tau)K(\tau)\right]^T H(\tau,r) - H(\tau,r)\left[A(\tau)+B(\tau)K(\tau)\right]$$
$$- \sum_{s=1}^{r-1}\frac{2r!}{s!(r-s)!}H(\tau,s)G(\tau)WG^T(\tau)H(\tau,r-s), \quad r \geq 2 \tag{2.21}$$

$$\frac{d}{d\tau}\breve{D}(\tau,1) = -\left[A(\tau)+B(\tau)K(\tau)\right]^T \breve{D}(\tau,1) - H(\tau,1)B(\tau)l(\tau)$$
$$- K^T(\tau)R(\tau)l(\tau) + C^T(\tau)Q(\tau)\zeta(\tau), \tag{2.22}$$

$$\frac{d}{d\tau}\breve{D}(\tau,r) = -\left[A(\tau)+B(\tau)K(\tau)\right]^T \breve{D}(\tau,r) - H(\tau,r)B(\tau)l(\tau), \quad r \geq 2 \tag{2.23}$$

and

$$\frac{d}{d\tau}D(\tau,1) = -\mathrm{Tr}\left\{H(\tau,1)G(\tau)WG^T(\tau)\right\} - 2\breve{D}^T(\tau,1)B(\tau)l(\tau)$$
$$- l^T(\tau)R(\tau)l(\tau) - \zeta^T(\tau)Q(\tau)\zeta(\tau), \tag{2.24}$$

$$\frac{d}{d\tau}D(\tau,r) = -\mathrm{Tr}\left\{H(\tau,r)G(\tau)WG^T(\tau)\right\} - 2\breve{D}^T(\tau,r)B(\tau)l(\tau),\ 2 \leq r \leq k, \tag{2.25}$$

where the terminal-value conditions $H(t_f,1) = C^T(t_f)Q_fC(t_f)$, $H(t_f,r) = 0$ *for* $2 \leq r \leq k$; $\breve{D}(t_f,1) = -C^T(t_f)Q_f\zeta(t_f)$, $\breve{D}(t_f,r) = 0$ *for* $2 \leq r \leq k$ *and* $D(t_f,1) = \zeta^T(t_f)Q_f\zeta(t_f)$, $D(t_f,r) = 0$ *for* $2 \leq r \leq k$.

Proof. Notice that the time-backward differential equations (2.20), (2.22) and (2.24) satisfied by $H(\tau,1)$, $\breve{D}(\tau,1)$, and $D(\tau,1)$ can be obtained by taking the derivative with respect to θ of the backward-in-time differential equations (2.11)–(2.13)

$$\frac{d}{d\tau}\left\{\frac{\partial}{\partial\theta}\Upsilon(\tau,\theta)\right\} = -\left[A(\tau)+B(\tau)K(\tau)\right]^T\frac{\partial}{\partial\theta}\Upsilon(\tau,\theta)$$
$$-\frac{\partial}{\partial\theta}\Upsilon(\tau,\theta)\left[A(\tau)+B(\tau)K(\tau)\right]$$

$$
\begin{aligned}
&-2\left\{\frac{\partial}{\partial\theta}\Upsilon(\tau,\theta)\right\}G(\tau)WG^T(\tau)\Upsilon(\tau,\theta)\\
&-2\Upsilon(\tau,\theta)G(\tau)WG^T(\tau)\left\{\frac{\partial}{\partial\theta}\Upsilon(\tau,\theta)\right\}\\
&-C^T(\tau)Q(\tau)C(\tau)-K^T(\tau)R(\tau)K(\tau),
\end{aligned}
$$

$$
\begin{aligned}
\frac{\mathrm{d}}{\mathrm{d}\tau}\left\{\frac{\partial}{\partial\theta}\eta(\tau,\theta)\right\} = &-[A(\tau)+B(\tau)K(\tau)]^T\frac{\partial}{\partial\theta}\eta(\tau,\theta)-\frac{\partial}{\partial\theta}\Upsilon(\tau,\theta)B(\tau)l(\tau)\\
&-K^T(\tau)R(\tau)l(\tau)+C^T(\tau)Q(\tau)\zeta(\tau),
\end{aligned}
$$

$$
\begin{aligned}
\frac{\mathrm{d}}{\mathrm{d}\tau}\left\{\frac{\partial}{\partial\theta}\upsilon(\tau,\theta)\right\} = &-\mathrm{Tr}\left\{\frac{\partial}{\partial\theta}\Upsilon(\tau,\theta)G(\tau)WG^T(\tau)\right\}-2\frac{\partial}{\partial\theta}\eta^T(\tau,\theta)B(\tau)l(\tau)\\
&-l^T(\tau)R(\tau)l(\tau)-\zeta^T(\tau)Q(\tau)\zeta(\tau),
\end{aligned}
$$

whereby the terminal-value conditions $\frac{\partial}{\partial\theta}\Upsilon(t_f,\theta)=C^T(t_f)Q_fC(t_f)$, $\frac{\partial}{\partial\theta}\eta(t_f,\theta)=-C^T(t_f)Q_f\zeta(t_f)$ and $\frac{\partial}{\partial\theta}\upsilon(t_f,\theta)=\zeta^T(t_f)Q_f\zeta(t_f)$. On the other hand, it is important to see that when $\theta=0$ the differential equation (2.11) becomes

$$
\begin{aligned}
\frac{\mathrm{d}}{\mathrm{d}\tau}\Upsilon(\tau,0) = &-[A(\tau)+B(\tau)K(\tau)]^T\Upsilon(\tau,0)-\Upsilon(\tau,0)[A(\tau)+B(\tau)K(\tau)]\\
&-2\Upsilon(\tau,0)G(\tau)WG^T(\tau)\Upsilon(\tau,0), \quad \Upsilon(t_f,0)=0.
\end{aligned}
$$

Broadly defined, the closed-loop matrix $A(\tau)+B(\tau)K(\tau)$ is assumed stable for all $\tau\in[t_0,t_f]$; it is therefore deduced that $\Upsilon(\tau,0)=0$. Using this result together with the definitions of $H(\tau,1)\triangleq\frac{\partial}{\partial\theta}\Upsilon(\tau,\theta)\Big|_{\theta=0}$, $\breve{D}(\tau,1)\triangleq\frac{\partial}{\partial\theta}\eta(\tau,\theta)\Big|_{\theta=0}$ and $D(\tau,1)\triangleq\frac{\partial}{\partial\theta}\upsilon(\tau,\theta)\Big|_{\theta=0}$, the first performance-measure statistic is found as follows:

$$
\kappa_1 = x_0^T H(t_0,1)x_0+2x_0^T\breve{D}(t_0,1)+D(t_0,1),
$$

whereby the solutions $H(\tau,1)$, $\breve{D}(\tau,1)$ and $D(\tau,1)$ satisfy the backward-in-time differential equations

$$
\begin{aligned}
\frac{\mathrm{d}}{\mathrm{d}\tau}H(\tau,1) = &-[A(\tau)+B(\tau)K(\tau)]^T H(\tau,1)-H(\tau,1)[A(\tau)+B(\tau)K(\tau)]\\
&-C^T(\tau)Q(\tau)C(\tau)-K^T(\tau)R(\tau)K(\tau),
\end{aligned}
$$

$$\frac{d}{d\tau}\breve{D}(\tau,1) = -[A(\tau)+B(\tau)K(\tau)]^T \breve{D}(\tau,1) - H(\tau,1)B(\tau)l(\tau)$$
$$- K^T(\tau)R(\tau)l(\tau) + C^T(\tau)Q(\tau)\zeta(\tau),$$
$$\frac{d}{d\tau}D(\tau,1) = -\mathrm{Tr}\left\{H(\tau,1)G(\tau)WG^T(\tau)\right\} - 2\breve{D}^T(\tau,1)B(\tau)l(\tau)$$
$$- l^T(\tau)R(\tau)l(\tau) - \zeta^T(\tau)Q(\tau)\zeta(\tau)$$

subject to the terminal values $H(t_f,1) = C^T(t_f)Q_fC(t_f)$, $\breve{D}(t_f,1) = -C^T(t_f)Q_f\zeta(t_f)$ and $D(t_f,1) = \zeta^T(t_f)Q_f\zeta(t_f)$.

Similarly, taking $\frac{\partial^2}{\partial\theta^2}$ of the time-backward differential equations (2.11)–(2.13) yield the corresponding differential equations

$$\frac{d}{d\tau}\left\{\frac{\partial^2}{\partial\theta^2}\Upsilon(\tau,\theta)\right\} = -[A(\tau)+B(\tau)K(\tau)]^T \frac{\partial^2}{\partial\theta^2}\Upsilon(\tau,\theta)$$
$$- \frac{\partial^2}{\partial\theta^2}\Upsilon(\tau,\theta)[A(\tau)+B(\tau)K(\tau)]$$
$$-2\frac{\partial^2}{\partial\theta^2}\Upsilon(\tau,\theta)G(\tau)WG^T(\tau)\Upsilon(\tau,\theta)$$
$$-4\frac{\partial}{\partial\theta}\Upsilon(\tau,\theta)G(\tau)WG^T(\tau)\frac{\partial}{\partial\theta}\Upsilon(\tau,\theta)$$
$$-2\Upsilon(\tau,\theta)G(\tau)WG^T(\tau)\frac{\partial^2}{\partial\theta^2}\Upsilon(\tau,\theta),$$

$$\frac{d}{d\tau}\left\{\frac{\partial^2}{\partial\theta^2}\eta(\tau,\theta)\right\} = -[A(\tau)+B(\tau)K(\tau)]^T\frac{\partial^2}{\partial\theta^2}\eta(\tau,\theta) - \frac{\partial^2}{\partial\theta^2}\Upsilon(\tau,\theta)B(\tau)l(\tau),$$
$$\frac{d}{d\tau}\left\{\frac{\partial^2}{\partial\theta^2}\upsilon(\tau,\theta)\right\} = -\mathrm{Tr}\left\{\frac{\partial^2}{\partial\theta^2}\Upsilon(\tau,\theta)G(\tau)WG^T(\tau)\right\} - 2\frac{\partial^2}{\partial\theta^2}\eta^T(\tau,\theta)B(\tau)l(\tau),$$

whereby the terminal values $\frac{\partial^2}{\partial\theta^2}\Upsilon(t_f,\theta) = 0$, $\frac{\partial^2}{\partial\theta^2}\eta(t_f,\theta) = 0$, and $\frac{\partial^2}{\partial\theta^2}\upsilon(t_f,\theta) = 0$.

Having replaced $H(\tau,1) \triangleq \frac{\partial}{\partial\theta}\Upsilon(\tau,\theta)\Big|_{\theta=0}$, $H(\tau,2) \triangleq \frac{\partial^2}{\partial\theta^2}\Upsilon(\tau,\theta)\Big|_{\theta=0}$, $\breve{D}(\tau,2) \triangleq \frac{\partial^2}{\partial\theta^2}\eta(\tau,\theta)\Big|_{\theta=0}$, $D(\tau,2) \triangleq \frac{\partial^2}{\partial\theta^2}\upsilon(\tau,\theta)\Big|_{\theta=0}$ and $\Upsilon(\tau,\theta)|_{\theta=0} = 0$ into the above equations, the second-order performance-measure statistics is now given by

$$\kappa_2 = x_0^T H(t_0,2)x_0 + 2x_0^T\breve{D}(t_0,2) + D(t_0,2)$$

in which the solutions $H(\tau,2)$, $\breve{D}(\tau,2)$, and $D(\tau,2)$ evaluated at $\tau = t_0$ are solving the time-backward differential differential equations

$$\frac{d}{d\tau}H(\tau,2) = -[A(\tau)+B(\tau)K(\tau)]^T H(\tau,2) - H(\tau,2)[A(\tau)+B(\tau)K(\tau)] \\ -4H(\tau,1)G(\tau)WG^T(\tau)H(\tau,1), \quad H(t_f,2)=0,$$

$$\frac{d}{d\tau}\breve{D}(\tau,2) = -[A(\tau)+B(\tau)K(\tau)]^T \breve{D}(\tau,2) - H(\tau,2)B(\tau)l(\tau), \quad \breve{D}(t_f,2)=0,$$

$$\frac{d}{d\tau}D(\tau,2) = -\mathrm{Tr}\left\{H(\tau,2)G(\tau)WG^T(\tau)\right\} - 2\breve{D}^T(\tau,2)B(\tau)l(\tau), \quad D(t_f,2)=0.$$

In general, having seen the basic procedure of generating the first two statistics of Eq. (2.7), the rth statistic associated with Eq. (2.7) is obtained by taking r-time derivatives of the backward-in-time differential equations (2.11)–(2.13) with respect to θ

$$\kappa_r = x_0^T H(t_0,r)x_0 + 2x_0^T \breve{D}(t_0,r) + D(t_0,r)$$

wherein $H(\tau,r)$, $\breve{D}(\tau,r)$ and $D(\tau,r)$ evaluated at $\tau = t_0$ are the solutions of the time-backward differential equations

$$\frac{d}{d\tau}H(\tau,r) = -[A(\tau)+B(\tau)K(\tau)]^T H(\tau,r) - H(\tau,r)[A(\tau)+B(\tau)K(\tau)] \\ -\sum_{s=1}^{r-1}\frac{2r!}{s!(r-s)!}H(\tau,s)G(\tau)WG^T(\tau)H(\tau,r-s), \quad H(t_f,r)=0,$$

$$\frac{d}{d\tau}\breve{D}(\tau,r) = -[A(\tau)+B(\tau)K(\tau)]^T \breve{D}(\tau,r) - H(\tau,r)B(\tau)l(\tau), \quad \breve{D}(t_f,r)=0,$$

$$\frac{d}{d\tau}D(\tau,r) = -\mathrm{Tr}\left\{H(\tau,r)G(\tau)WG^T(\tau)\right\} - 2\breve{D}^T(\tau,r)B(\tau)l(\tau), \quad D(t_f,r)=0.$$

□

2.4 Statements of Mayer Problem with Performance Risk Aversion

Rigorously speaking, risk-value aware performance appraisal which is a multi-objective criterion of performance-measure statistics of the random cost (2.7) by risk-averse control designers has both forward-and backward-looking aspects in terms of time. It is forward looking in the sense that risk-value aware performance appraisal provides information for possible direct intervention, screening of risk-averse attitudes, and revision of control as corrective actions for performance risk aversion. On the other hand, risk-value performance appraisal is backward looking because the existence of risk-value aware performance appraisal ex post facto in many instances affects the controller's behavior which has produced the past performance that risk-averse control designers try to appraise. In this sense, performance-measure statistics affect the controller's behavior by influencing the consequences of its actions in terms of implicit or explicit performance values and risks.

As have been emphasized all along, the basic reason why it is important to consider risk-averse control behavior is the presence of performance uncertainty. Therefore, the dynamics (2.20)–(2.25) of performance-measure statistics (2.19) are the focus of attention. In preparing for the statements of statistical optimal control problem with tracking applications herein problem, let the k-tuple state variables $\mathscr{H}$, $\breve{\mathscr{D}}$ and $\mathscr{D}$ be defined as follows: $\mathscr{H}(\cdot) \triangleq (\mathscr{H}_1(\cdot),\ldots,\mathscr{H}_k(\cdot))$, $\breve{\mathscr{D}}(\cdot) \triangleq (\breve{\mathscr{D}}_1(\cdot),\ldots,\breve{\mathscr{D}}_k(\cdot))$ and $\mathscr{D}(\cdot) \triangleq (\mathscr{D}_1(\cdot),\ldots,\mathscr{D}_k(\cdot))$ whereby each element $\mathscr{H}_r \in \mathscr{C}^1([t_0,t_f];\mathbb{R}^{n\times n})$ of $\mathscr{H}$, $\breve{\mathscr{D}}_r \in \mathscr{C}^1([t_0,t_f];\mathbb{R}^n)$ of $\breve{\mathscr{D}}$ and $\mathscr{D}_r \in \mathscr{C}^1([t_0,t_f];\mathbb{R})$ of $\mathscr{D}$ has the representations

$$\mathscr{H}_r(\cdot) = H(\cdot,r), \qquad \breve{\mathscr{D}}_r(\cdot) = \breve{D}(\cdot,r), \qquad \mathscr{D}_r(\cdot) = D(\cdot,r)$$

with the right members satisfying the dynamic equations (2.20)–(2.25) on the finite horizon $[t_0,t_f]$. The problem formulation can be considerably simplified if the bounded and Lipschitz continuous mappings are introduced

$$\mathscr{F}_r : [t_0,t_f] \times (\mathbb{R}^{n\times n})^k \times \mathbb{R}^{m\times n} \mapsto \mathbb{R}^{n\times n},$$

$$\breve{\mathscr{G}}_r : [t_0,t_f] \times (\mathbb{R}^{n\times n})^k \times (\mathbb{R}^n)^k \times \mathbb{R}^{m\times n} \times \mathbb{R}^m \mapsto \mathbb{R}^n,$$

$$\mathscr{G}_r : [t_0,t_f] \times (\mathbb{R}^{n\times n})^k \times (\mathbb{R}^n)^k \times \mathbb{R}^m \mapsto \mathbb{R}$$

where the actions are given by

$$\begin{aligned}
\mathscr{F}_1(\tau,\mathscr{H},K) &= -[A(\tau)+B(\tau)K(\tau)]^T\, \mathscr{H}_1(\tau) - \mathscr{H}_1(\tau)\,[A(\tau)+B(\tau)K(\tau)] \\
&\quad - C^T(\tau)Q(\tau)C(\tau) - K^T(\tau)R(\tau)K(\tau), \\
\mathscr{F}_r(\tau,\mathscr{H},K) &= -[A(\tau)+B(\tau)K(\tau)]^T\, \mathscr{H}_r(\tau) - \mathscr{H}_r(\tau)\,[A(\tau)+B(\tau)K(\tau)] \\
&\quad - \sum_{s=1}^{r-1} \frac{2r!}{s!(r-s)!} \mathscr{H}_s(\tau)G(\tau)WG^T(\tau)\mathscr{H}_{r-s}(\tau),
\end{aligned}$$

$$\begin{aligned}
\breve{\mathscr{G}}_1(\tau,\mathscr{H},\breve{\mathscr{D}},K,l) &= -[A(\tau)+B(\tau)K(\tau)]^T\, \breve{\mathscr{D}}_1(\tau) - \mathscr{H}_1(\tau)B(\tau)l(\tau) \\
&\quad - K^T(\tau)R(\tau)l(\tau) + C^T(\tau)Q(\tau)\zeta(\tau), \\
\breve{\mathscr{G}}_r(\tau,\mathscr{H},\breve{\mathscr{D}},K,l) &= -[A(\tau)+B(\tau)K(\tau)]^T\, \breve{\mathscr{D}}_r(\tau) - \mathscr{H}_r(\tau)B(\tau)l(\tau),
\end{aligned}$$

$$\mathscr{G}_1\left(\tau,\mathscr{H},\breve{\mathscr{D}},l\right) = -\mathrm{Tr}\left\{\mathscr{H}_1(\tau)G(\tau)WG^T(\tau)\right\} - 2\breve{\mathscr{D}}_1^T(\tau)B(\tau)l(\tau) - l^T(\tau)R(\tau)l(\tau) - \zeta^T(\tau)Q(\tau)\zeta(\tau),$$

$$\mathscr{G}_r\left(\tau,\mathscr{H},\breve{\mathscr{D}},l\right) = -\mathrm{Tr}\left\{\mathscr{H}_r(\tau)G(\tau)WG^T(\tau)\right\} - 2\breve{\mathscr{D}}_r^T(\tau)B(\tau)l(\tau).$$

Now there is no difficulty to establish the product mappings

$$\mathscr{F}_1\times\cdots\times\mathscr{F}_k : [t_0,t_f]\times(\mathbb{R}^{n\times n})^k\times\mathbb{R}^{m\times n}\mapsto(\mathbb{R}^{n\times n})^k,$$
$$\breve{\mathscr{G}}_1\times\cdots\times\breve{\mathscr{G}}_k : [t_0,t_f]\times(\mathbb{R}^{n\times n})^k\times(\mathbb{R}^n)^k\times\mathbb{R}^{m\times n}\times\mathbb{R}^m\mapsto(\mathbb{R}^n)^k,$$
$$\mathscr{G}_1\times\cdots\times\mathscr{G}_k : [t_0,t_f]\times(\mathbb{R}^{n\times n})^k\times(\mathbb{R}^n)^k\times\mathbb{R}^m\mapsto\mathbb{R}^k$$

along with the corresponding notations $\mathscr{F}=\mathscr{F}_1\times\cdots\times\mathscr{F}_k$, $\breve{\mathscr{G}}=\breve{\mathscr{G}}_1\times\cdots\times\breve{\mathscr{G}}_k$, and $\mathscr{G}=\mathscr{G}_1\times\cdots\times\mathscr{G}_k$. Thus, the dynamic equations of motion (2.20)–(2.25) can be rewritten as

$$\frac{\mathrm{d}}{\mathrm{d}\tau}\mathscr{H}(\tau)=\mathscr{F}(\tau,\mathscr{H}(\tau),K(\tau)),\qquad \mathscr{H}(t_f)=\mathscr{H}_f,$$
$$\frac{\mathrm{d}}{\mathrm{d}\tau}\breve{\mathscr{D}}(\tau)=\breve{\mathscr{G}}\left(\tau,\mathscr{H}(\tau),\breve{\mathscr{D}}(\tau),K(\tau),l(\tau)\right),\qquad \breve{\mathscr{D}}(t_f)=\breve{\mathscr{D}}_f,$$
$$\frac{\mathrm{d}}{\mathrm{d}\tau}\mathscr{D}(\tau)=\mathscr{G}\left(\tau,\mathscr{H}(\tau),\breve{\mathscr{D}}(\tau),l(\tau)\right),\qquad \mathscr{D}(t_f)=\mathscr{D}_f,$$

whereby the k-tuple terminal-value conditions $\mathscr{H}_f=\left(C^T(t_f)Q_fC(t_f),0,\ldots,0\right)$, $\breve{\mathscr{D}}_f=\left(-C^T(t_f)Q_f\zeta(t_f),0,\ldots,0\right)$ and $\mathscr{D}_f=(0,\ldots,0)$.

Notice that the product system uniquely determines $\mathscr{H}$, $\breve{\mathscr{D}}$, and $\mathscr{D}$ once the admissible affine input l and feedback gain K are specified. Henceforth, they are considered as $\mathscr{H}=\mathscr{H}(\cdot,K)$, $\breve{\mathscr{D}}=\breve{\mathscr{D}}(\cdot,K,l)$ and $\mathscr{D}=\mathscr{D}(\cdot,K,l)$.

Next, the effects of what are considered the degrees of performance uncertainty are in terms of performance-measure statistics beyond the expected level of performance measure (2.7). The greater the higher-order performance-measure statistics, the greater the degrees of performance uncertainty or performance risks. Similarly, the degrees of uncertainty of one probability distribution of a random performance measure are greater than another probability distribution if the linear combination of first finite numbers of performance-measure statistics of the first distribution is greater than that of the second distribution. The following performance appraisal therefore addresses the concern of performance value and risk awareness:

Definition 2.4.1 (Risk-Value Aware Performance Index). Let $k\in\mathbb{Z}^+$ and the sequence $\mu=\{\mu_r\geq 0\}_{r=1}^k$ with $\mu_1>0$. Then, for the admissible (t_0,x_0), the performance index with risk and value awareness

$$\phi_0^{tk} : [t_0,t_f]\times(\mathbb{R}^{n\times n})^k\times(\mathbb{R}^n)^k\times\mathbb{R}^k\mapsto\mathbb{R}^+$$

in statistical optimal control on $[t_0,t_f]$ is of Mayer type and defined by

$$\phi_0^{tk}\left(t_0,\mathscr{H}(t_0),\breve{\mathscr{D}}(t_0),\mathscr{D}(t_0)\right) \triangleq \underbrace{\mu_1\kappa_1}_{\text{Value Measure}} \quad \underbrace{+\mu_2\kappa_2+\cdots+\mu_k\kappa_k}_{\text{Risk Measures}}$$

$$= \sum_{r=1}^{k} \mu_r \left[x_0^T \mathscr{H}_r(t_0)x_0 + 2x_0^T \breve{\mathscr{D}}_r(t_0) + \mathscr{D}_r(t_0)\right], \quad (2.26)$$

where the scalar, real constants μ_r represent parametric design of freedom and the unique solutions $\{\mathscr{H}_r(t_0) \geq 0\}_{r=1}^k$, $\{\breve{\mathscr{D}}_r(t_0)\}_{r=1}^k$ and $\{\mathscr{D}_r(t_0)\}_{r=1}^k$ evaluated at $\tau = t_0$ satisfy the dynamical equations

$$\frac{\mathrm{d}}{\mathrm{d}\tau}\mathscr{H}(\tau) = \mathscr{F}(\tau,\mathscr{H}(\tau),K(\tau)), \quad \mathscr{H}(t_f) = \mathscr{H}_f,$$

$$\frac{\mathrm{d}}{\mathrm{d}\tau}\breve{\mathscr{D}}(\tau) = \breve{\mathscr{G}}\left(\tau,\mathscr{H}(\tau),\breve{\mathscr{D}}(\tau),K(\tau),l(\tau)\right), \quad \breve{\mathscr{D}}(t_f) = \breve{\mathscr{D}}_f,$$

$$\frac{\mathrm{d}}{\mathrm{d}\tau}\mathscr{D}(\tau) = \mathscr{G}\left(\tau,\mathscr{H}(\tau),\breve{\mathscr{D}}(\tau),l(\tau)\right), \quad \mathscr{D}(t_f) = \mathscr{D}_f.$$

For the given terminal data $(t_f,\mathscr{H}_f,\breve{\mathscr{D}}_f,\mathscr{D}_f)$, the classes of admissible affine inputs and feedback gains that can be adjusted optimally to cope with the increased uncertainty of Eq. (2.7)are defined as follows.

Definition 2.4.2 (Admissible Affine Inputs and Feedback Gains). Let compact subsets $\overline{L} \subset \mathbb{R}^m$ and $\overline{K} \subset \mathbb{R}^{m\times n}$ be the sets of allowable affine inputs and gain values. For the given $k \in \mathbb{Z}^+$ and the sequence $\mu = \{\mu_r \geq 0\}_{r=1}^k$ with $\mu_1 > 0$, the set of admissible affine inputs $\mathscr{L}_{t_f,\mathscr{H}_f,\breve{\mathscr{D}}_f,\mathscr{D}_f;\mu}$ and feedback gains $\mathscr{K}_{t_f,\mathscr{H}_f,\breve{\mathscr{D}}_f,\mathscr{D}_f;\mu}$ are respectively assumed to be the classes of $\mathscr{C}([t_0,t_f];\mathbb{R}^m)$ and $\mathscr{C}([t_0,t_f];\mathbb{R}^{m\times n})$ with values $l(\cdot) \in \overline{L}$ and $K(\cdot) \in \overline{K}$ for which the solutions to the dynamic equations with the terminal-value conditions $\mathscr{H}(t_f) = \mathscr{H}_f$, $\breve{\mathscr{D}}(t_f) = \breve{\mathscr{D}}_f$ and $\mathscr{D}(t_f) = \mathscr{D}_f$

$$\frac{\mathrm{d}}{\mathrm{d}\tau}\mathscr{H}(\tau) = \mathscr{F}(\tau,\mathscr{H}(\tau),K(\tau)), \quad (2.27)$$

$$\frac{\mathrm{d}}{\mathrm{d}\tau}\breve{\mathscr{D}}(\tau) = \breve{\mathscr{G}}\left(\tau,\mathscr{H}(\tau),\breve{\mathscr{D}}(\tau),K(\tau),l(\tau)\right), \quad (2.28)$$

$$\frac{\mathrm{d}}{\mathrm{d}\tau}\mathscr{D}(\tau) = \mathscr{G}\left(\tau,\mathscr{H}(\tau),\breve{\mathscr{D}}(\tau),l(\tau)\right) \quad (2.29)$$

exist on the interval of optimization $[t_0,t_f]$.

Next the optimization statements for the working of linear-quadratic tracking, especially one conducive to appropriate risk-averse behavior over a finite horizon are stated in the sequel.

Definition 2.4.3 (Optimization Problem of Mayer Type). Fix $k \in \mathbb{Z}^+$ and the sequence $\mu = \{\mu_r \geq 0\}_{r=1}^k$ with $\mu_1 > 0$. Then, the Mayer optimization problem over $[t_0, t_f]$ is given by the minimization of the risk-value aware performance index (2.26) over $l(\cdot) \in \mathscr{L}_{t_f,\mathscr{H}_f,\breve{\mathscr{D}}_f,\mathscr{D}_f;\mu}$, $K(\cdot) \in \mathscr{K}_{t_f,\mathscr{H}_f,\breve{\mathscr{D}}_f,\mathscr{D}_f;\mu}$ and subject to the dynamic equations of motion (2.27)–(2.29) for $\tau \in [t_0, t_f]$.

Construction of scalar-valued functions which are the candidates for the value function plays a key role in the dynamic programming approach and leads directly to the concept of a reachable set in the latter half of this chapter.

Definition 2.4.4 (Reachable Set). Let reachable set $\mathscr{Q}$ be defined $\mathscr{Q} \triangleq \{(\varepsilon, \mathscr{Y}, \breve{\mathscr{Z}}, \mathscr{Z}) \in [t_0, t_f] \times (\mathbb{R}^{n\times n})^k \times (\mathbb{R}^n)^k \times \mathbb{R}^k\}$ such that $\mathscr{L}_{\varepsilon,\mathscr{Y},\breve{\mathscr{Z}},\mathscr{Z};\mu} \times \mathscr{K}_{\varepsilon,\mathscr{Y},\breve{\mathscr{Z}},\mathscr{Z};\mu} \neq \emptyset$.

By adapting to the initial cost problem and the terminologies present in the statistical optimal control here, the Hamilton-Jacobi-Bellman (HJB) equation satisfied by the value function $\mathscr{V}(\varepsilon, \mathscr{Y}, \breve{\mathscr{Z}}, \mathscr{Z})$ is then given as follows.

Theorem 2.4.1 (HJB Equation for Mayer Problem). *Let $(\varepsilon, \mathscr{Y}, \breve{\mathscr{Z}}, \mathscr{Z})$ be any interior point of the reachable set $\mathscr{Q}$ at which the value function $\mathscr{V}(\varepsilon, \mathscr{Y}, \breve{\mathscr{Z}}, \mathscr{Z})$ is differentiable. If there exist optimal affine input $l^* \in \mathscr{L}_{\varepsilon,\mathscr{Y},\breve{\mathscr{Z}},\mathscr{Z};\mu}$ and feedback gain $K^* \in \mathscr{K}_{\varepsilon,\mathscr{Y},\breve{\mathscr{Z}},\mathscr{Z};\mu}$, then the partial differential equation of dynamic programming*

$$
\begin{aligned}
0 = \min_{l\in\overline{L},K\in\overline{K}} \Big\{ & \frac{\partial}{\partial\varepsilon}\mathscr{V}(\varepsilon, \mathscr{Y}, \breve{\mathscr{Z}}, \mathscr{Z}) + \frac{\partial}{\partial \operatorname{vec}(\mathscr{Y})}\mathscr{V}(\varepsilon, \mathscr{Y}, \breve{\mathscr{Z}}, \mathscr{Z}) \operatorname{vec}(\mathscr{F}(\varepsilon, \mathscr{Y}, K)) \\
& + \frac{\partial}{\partial \operatorname{vec}(\breve{\mathscr{Z}})}\mathscr{V}(\varepsilon, \mathscr{Y}, \breve{\mathscr{Z}}, \mathscr{Z}) \operatorname{vec}(\breve{\mathscr{G}}(\varepsilon, \mathscr{Y}, \breve{\mathscr{Z}}, K, l)) \\
& + \frac{\partial}{\partial \operatorname{vec}(\mathscr{Z})}\mathscr{V}(\varepsilon, \mathscr{Y}, \breve{\mathscr{Z}}, \mathscr{Z}) \operatorname{vec}(\mathscr{G}(\varepsilon, \mathscr{Y}, \breve{\mathscr{Z}}, l)) \Big\}
\end{aligned} \tag{2.30}
$$

is satisfied when the boundary condition is given by

$$
\mathscr{V}(t_0, \mathscr{H}(t_0), \breve{\mathscr{D}}(t_0), \mathscr{D}(t_0)) = \phi_0^{tk}(t_0, \mathscr{H}(t_0), \breve{\mathscr{D}}(t_0), \mathscr{D}(t_0)) .
$$

Proof. The proof can be obtained by adapting the results from [18] with the aid of the isomorphic mapping $\operatorname{vec}(\cdot)$ from the whole vector space of $\mathbb{R}^{n_1\times n_2}$ to the entire vector space of $\mathbb{R}^{n_1 n_2}$ as shown in [20]. □

Theorem 2.4.2 (Verification Theorem). *Fix $k \in \mathbb{Z}^+$ and let $\mathscr{W}(\varepsilon, \mathscr{Y}, \breve{\mathscr{Z}}, \mathscr{Z})$ be a continuously differentiable solution of the HJB equation (2.30) which satisfies the boundary condition*

$$
\mathscr{W}(t_0, \mathscr{H}(t_0), \breve{\mathscr{D}}(t_0), \mathscr{D}(t_0)) = \phi_0^{tk}(t_0, \mathscr{H}(t_0), \breve{\mathscr{D}}(t_0), \mathscr{D}(t_0)) . \tag{2.31}
$$

Let $(t_f, \mathscr{H}_f, \breve{\mathscr{D}}_f, \mathscr{D}_f)$ be in $\mathscr{Q}$; (l, K) in $\mathscr{L}_{t_f,\mathscr{H}_f,\breve{\mathscr{D}}_f,\mathscr{D}_f;\mu} \times \mathscr{K}_{t_f,\mathscr{H}_f,\breve{\mathscr{D}}_f,\mathscr{D}_f;\mu}$; $\mathscr{H}$, $\breve{\mathscr{D}}$ and $\mathscr{D}$ the corresponding solutions of Eqs. (2.27)–(2.29). Then

$\mathscr{W}(\tau,\mathscr{H}(\tau),\breve{\mathscr{D}}(\tau),\mathscr{D}(\tau))$ is a non-increasing function of τ. If the 2-tuple (l^*,K^*) is in $\mathscr{L}_{t_f,\mathscr{H}_f,\breve{\mathscr{D}}_f,\mathscr{D}_f;\mu} \times \mathscr{K}_{t_f,\mathscr{H}_f,\breve{\mathscr{D}}_f,\mathscr{D}_f;\mu}$ defined on $[t_0,t_f]$ with corresponding solutions, $\mathscr{H}^*$, $\breve{\mathscr{D}}^*$ and $\mathscr{D}^*$ of the dynamical equations (2.27)–(2.29) such that for $\tau \in [t_0,t_f]$

$$\begin{aligned}
0 &= \frac{\partial}{\partial \varepsilon}\mathscr{W}\left(\tau,\mathscr{H}^*(\tau),\breve{\mathscr{D}}^*(\tau),\mathscr{D}^*(\tau)\right) \\
&+\frac{\partial}{\partial \mathrm{vec}(\mathscr{Y})}\mathscr{W}\left(\tau,\mathscr{H}^*(\tau),\breve{\mathscr{D}}^*(\tau),\mathscr{D}^*(\tau)\right)\mathrm{vec}\left(\mathscr{F}\left(\tau,\mathscr{H}^*(\tau),K^*(\tau)\right)\right) \\
&+\frac{\partial}{\partial \mathrm{vec}(\breve{\mathscr{Z}})}\mathscr{W}\left(\tau,\mathscr{H}^*(\tau),\breve{\mathscr{D}}^*(\tau),\mathscr{D}^*(\tau)\right)\mathrm{vec}\left(\breve{\mathscr{G}}\left(\tau,\mathscr{H}^*(\tau),\breve{\mathscr{D}}^*(\tau),K^*(\tau),l^*(\tau)\right)\right) \\
&+\frac{\partial}{\partial \mathrm{vec}(\mathscr{Z})}\mathscr{W}\left(\tau,\mathscr{H}^*(\tau),\breve{\mathscr{D}}^*(\tau),\mathscr{D}^*(\tau)\right)\mathrm{vec}\left(\mathscr{G}\left(\tau,\mathscr{H}^*(\tau),\breve{\mathscr{D}}^*(\tau),l^*(\tau)\right)\right)
\end{aligned} \tag{2.32}$$

then l^* and K^* are optimal. Moreover,

$$\mathscr{W}\left(\varepsilon,\mathscr{Y},\breve{\mathscr{Z}},\mathscr{Z}\right) = \mathscr{V}\left(\varepsilon,\mathscr{Y},\breve{\mathscr{Z}},\mathscr{Z}\right) \tag{2.33}$$

where $\mathscr{V}\left(\varepsilon,\mathscr{Y},\breve{\mathscr{Z}},\mathscr{Z}\right)$ is the value function.

Proof. The proof is relegated to the adaptation of the Mayer-form verification theorem in [18]. And the technical details can be found in [20]. □

2.5 Risk-Averse Control of Adaptive Behavior

In this section, a closed-loop control solution of risk-averse adaptive behavior within a finite horizon of the statistical optimal control problem is presented. Next, the terminal time and states $\left(t_f,\mathscr{H}_f,\breve{\mathscr{D}}_f,\mathscr{D}_f\right)$ of the dynamical equations are generalized to $\left(\varepsilon,\mathscr{Y},\breve{\mathscr{Z}},\mathscr{Z}\right)$ in the dynamic programming framework. That is, for $\varepsilon \in [t_0,t_f]$ and $1 \le r \le k$, the states of the dynamical system (2.27)–(2.29) defined on the interval $[t_0,\varepsilon]$ have the terminal values denoted by

$$\mathscr{H}(\varepsilon) = \mathscr{Y}\,, \qquad \breve{\mathscr{D}}(\varepsilon) = \breve{\mathscr{Z}}\,, \qquad \mathscr{D}(\varepsilon) = \mathscr{Z}\,.$$

The mathematical property of the risk-value aware performance index (2.26) is quadratic affine in terms of the arbitrary initial system state x_0. Therefore, it is used to interpret in finding a solution to the HJB equation (2.30) as follows:

$$
\begin{aligned}
\mathscr{W}\left(\varepsilon,\mathscr{Y},\breve{\mathscr{Z}},\mathscr{Z}\right) &= x_0^T \sum_{r=1}^{k} \mu_r \left(\mathscr{Y}_r + \mathscr{E}_r(\varepsilon)\right) x_0 \\
&\quad + 2x_0^T \sum_{r=1}^{k} \mu_r \left(\breve{\mathscr{Z}}_r + \breve{\mathscr{T}}_r(\varepsilon)\right) + \sum_{r=1}^{k} \mu_r \left(\mathscr{Z}_r + \mathscr{T}_r(\varepsilon)\right) \qquad (2.34)
\end{aligned}
$$

whereby the time parametric functions $\mathscr{E}_r \in \mathscr{C}^1([t_0,t_f];\mathbb{R}^{n\times n})$, $\breve{\mathscr{T}}_r \in \mathscr{C}^1([t_0,t_f];\mathbb{R}^n)$, and $\mathscr{T}_r \in \mathscr{C}^1([t_0,t_f];\mathbb{R})$ are to be determined. Using the isomorphic vec mapping, there is no difficulty to verify the following result.

Theorem 2.5.1. *Fix $k \in \mathbb{Z}^+$ and let $\left(\varepsilon,\mathscr{Y},\breve{\mathscr{Z}},\mathscr{Z}\right)$ be any interior point of the reachable set $\mathscr{Q}$ at which the real-valued function $\mathscr{W}\left(\varepsilon,\mathscr{Y},\breve{\mathscr{Z}},\mathscr{Z}\right)$ of the form (2.34) is differentiable. The derivative of $\mathscr{W}\left(\varepsilon,\mathscr{Y},\breve{\mathscr{Z}},\mathscr{Z}\right)$ with respect to ε is given*

$$
\begin{aligned}
\frac{d}{d\varepsilon}\mathscr{W}\left(\varepsilon,\mathscr{Y},\breve{\mathscr{Z}},\mathscr{Z}\right) &= x_0^T \sum_{r=1}^{k} \mu_r(\mathscr{F}_r(\varepsilon,\mathscr{Y},K) + \frac{d}{d\varepsilon}\mathscr{E}_r(\varepsilon))x_0 \\
&\quad + 2x_0^T \sum_{r=1}^{k} \mu_r(\breve{\mathscr{G}}_r\left(\varepsilon,\mathscr{Y},\breve{\mathscr{Z}},K,l\right) + \frac{d}{d\varepsilon}\breve{\mathscr{T}}_r(\varepsilon)) \\
&\quad + \sum_{r=1}^{k} \mu_r(\mathscr{G}_r\left(\varepsilon,\mathscr{Y},\breve{\mathscr{Z}},l\right) + \frac{d}{d\varepsilon}\mathscr{T}_r(\varepsilon)) \qquad (2.35)
\end{aligned}
$$

provided $l \in \overline{L}$ and $K \in \overline{K}$.

Proof. The approach taken is similar to that of [19] and also in [20]. □

Trying the guess solution (2.34) into the HJB equation for the Mayer problem here (2.30), it follows that

$$
\begin{aligned}
0 = \min_{l\in\overline{L},\,K\in\overline{K}} \Bigg\{ & x_0^T \sum_{r=1}^{k} \mu_r(\mathscr{F}_r(\varepsilon,\mathscr{Y},K) + \frac{d}{d\varepsilon}\mathscr{E}_r(\varepsilon))x_0 \\
& + 2x_0^T \sum_{r=1}^{k} \mu_r(\breve{\mathscr{G}}_r\left(\varepsilon,\mathscr{Y},\breve{\mathscr{Z}},K,l\right) + \frac{d}{d\varepsilon}\breve{\mathscr{T}}_r(\varepsilon)) \\
& + \sum_{r=1}^{k} \mu_r(\mathscr{G}_r\left(\varepsilon,\mathscr{Y},\breve{\mathscr{Z}},l\right) + \frac{d}{d\varepsilon}\mathscr{T}_r(\varepsilon)) \Bigg\}. \qquad (2.36)
\end{aligned}
$$

Notice that

$$\sum_{r=1}^{k}\mu_r\mathscr{F}_r(\varepsilon,\mathscr{Y},K) = -[A(\varepsilon)+B(\varepsilon)K]^T\sum_{r=1}^{k}\mu_r\mathscr{Y}_r - \sum_{r=1}^{k}\mu_r\mathscr{Y}_r[A(\varepsilon)+B(\varepsilon)K]$$
$$-\mu_1 C^T(\varepsilon)Q(\varepsilon)C(\varepsilon) - \mu_1 K^T R(\varepsilon)K$$
$$-\sum_{r=2}^{k}\mu_r\sum_{s=1}^{r-1}\frac{2r!}{s!(r-s)!}\mathscr{Y}_s G(\varepsilon)WG^T(\varepsilon)\mathscr{Y}_{r-s}$$

$$\sum_{r=1}^{k}\mu_r\breve{\mathscr{G}}_r(\varepsilon,\mathscr{Y},\breve{\mathscr{Z}},K,l) = -[A(\varepsilon)+B(\varepsilon)K]^T\sum_{r=1}^{k}\mu_r\breve{\mathscr{Z}}_r$$
$$-\sum_{r=1}^{k}\mu_r\mathscr{Y}_r B(\varepsilon)l - \mu_1 K^T R(\varepsilon)l + \mu_1 C^T(\varepsilon)Q(\varepsilon)\zeta(\varepsilon),$$

$$\sum_{r=1}^{k}\mu_r\mathscr{G}_r(\varepsilon,\mathscr{Y},\breve{\mathscr{Z}},l) = -\sum_{r=1}^{k}\mu_r\mathrm{Tr}\{\mathscr{Y}_r G(\varepsilon)WG^T(\varepsilon)\}$$
$$-2\sum_{r=1}^{k}\mu_r\breve{\mathscr{Z}}_r^T B(\varepsilon)l - \mu_1 l^T R(\varepsilon)l - \mu_1\zeta^T(\varepsilon)Q(\varepsilon)\zeta(\varepsilon).$$

Since the initial conditions x_0 and M_0 are arbitrary vector and rank-one matrix, the necessary condition for an extremum of Eq. (2.26) on $[t_0,\varepsilon]$ is obtained by differentiating the expression within the bracket of Eq. (2.36) with respect to l and K as

$$l(\varepsilon,\breve{\mathscr{Z}}) = -R^{-1}(\varepsilon)B^T(\varepsilon)\sum_{r=1}^{k}\hat{\mu}_r\breve{\mathscr{Z}}_r, \tag{2.37}$$

$$K(\varepsilon,\mathscr{Y}) = -R^{-1}(\varepsilon)B^T(\varepsilon)\sum_{r=1}^{k}\hat{\mu}_r\mathscr{Y}_r, \tag{2.38}$$

whereby the adjusted degrees of freedom $\hat{\mu}_r \triangleq \mu_r/\mu_1$ and $\mu_1 > 0$. Replacing the extremal affine input (2.37) and feedback gain (2.38) into (2.36) leads to the value function

$$x_0^T\Bigg[\sum_{r=1}^{k}\mu_r\frac{\mathrm{d}}{\mathrm{d}\varepsilon}\mathscr{E}_r(\varepsilon) - A^T(\varepsilon)\sum_{r=1}^{k}\mu_r\mathscr{Y}_r - \sum_{r=1}^{k}\mu_r\mathscr{Y}_r A(\varepsilon) - \mu_1 C^T(\varepsilon)Q(\varepsilon)C(\varepsilon)$$
$$+\sum_{r=1}^{k}\hat{\mu}_r\mathscr{Y}_r B(\varepsilon)R^{-1}(\varepsilon)B^T(\varepsilon)\sum_{s=1}^{k}\mu_s\mathscr{Y}_s + \sum_{s=1}^{k}\mu_s\mathscr{Y}_s(\varepsilon)B(\varepsilon)R^{-1}(\varepsilon)B^T(\varepsilon)\sum_{r=1}^{k}\hat{\mu}_r\mathscr{Y}_r$$
$$-\mu_1\sum_{r=1}^{k}\hat{\mu}_r\mathscr{Y}_r B(\varepsilon)R^{-1}(\varepsilon)B^T(\varepsilon)\sum_{s=1}^{k}\hat{\mu}_s\mathscr{Y}_s - \sum_{r=2}^{k}\mu_r\sum_{s=1}^{r-1}\frac{2r!}{s!(r-s)!}\mathscr{Y}_s G(\varepsilon)WG^T(\varepsilon)\mathscr{Y}_{r-s}\Bigg]x_0$$

$$
\begin{aligned}
&+2x_0^T\left[\sum_{r=1}^{k}\mu_r\frac{\mathrm{d}}{\mathrm{d}\varepsilon}\breve{\mathscr{T}}_r(\varepsilon)-A^T(\varepsilon)\sum_{r=1}^{k}\mu_r\breve{\mathscr{Z}}_r+\mu_1C^T(\varepsilon)Q(\varepsilon)\zeta(\varepsilon)\right.\\
&+\sum_{r=1}^{k}\mu_r\mathscr{Y}_rB(\varepsilon)R^{-1}(\varepsilon)B^T(\varepsilon)\sum_{s=1}^{k}\mu_s\breve{\mathscr{Z}}_s+\sum_{s=1}^{k}\mu_s\mathscr{Y}_sB(\varepsilon)R^{-1}(\varepsilon)B^T(\varepsilon)\sum_{r=1}^{k}\hat{\mu}_r\breve{\mathscr{Z}}_r\\
&\left.-\mu_1\sum_{r=1}^{k}\hat{\mu}_r\mathscr{Y}_rB(\varepsilon)R^{-1}(\varepsilon)B^T(\varepsilon)\sum_{s=1}^{k}\hat{\mu}_s\breve{\mathscr{Z}}_s\right]+\sum_{r=1}^{k}\mu_r\frac{\mathrm{d}}{\mathrm{d}\varepsilon}\mathscr{T}_r(\varepsilon)\\
&-\sum_{r=1}^{k}\mu_r\mathrm{Tr}\left\{\mathscr{Y}_rG(\varepsilon)WG^T(\varepsilon)\right\}+2\sum_{r=1}^{k}\mu_r\breve{\mathscr{Z}}_r^TB(\varepsilon)R^{-1}(\varepsilon)B^T(\varepsilon)\sum_{s=1}^{k}\hat{\mu}_s\breve{\mathscr{Z}}_s\\
&-\mu_1\zeta^T(\varepsilon)Q(\varepsilon)\zeta(\varepsilon)-\mu_1\sum_{r=1}^{k}\hat{\mu}_r\breve{\mathscr{Z}}_r^TB(\varepsilon)R^{-1}(\varepsilon)B^T(\varepsilon)\sum_{s=1}^{k}\hat{\mu}_s\breve{\mathscr{Z}}_s. \qquad (2.39)
\end{aligned}
$$

The remaining task is to display time-dependent functions $\{\mathscr{E}_r(\cdot)\}_{r=1}^k$, $\{\breve{\mathscr{T}}_r(\cdot)\}_{r=1}^k$ and $\{\mathscr{T}_r(\cdot)\}_{r=1}^k$, which yield a sufficient condition to have the left-hand side of Eq. (2.39) being zero for any $\varepsilon \in [t_0, t_f]$, when $\{\mathscr{Y}_r\}_{r=1}^k$, $\{\breve{\mathscr{Z}}_r\}_{r=1}^k$ and $\{\mathscr{Z}_r\}_{r=1}^k$ are evaluated along solutions to the dynamical equations (2.27)–(2.29). Careful observation of Eq. (2.39) suggests that $\{\mathscr{E}_r(\cdot)\}_{r=1}^k$, $\{\breve{\mathscr{T}}_r(\cdot)\}_{r=1}^k$, and $\{\mathscr{T}_r(\cdot)\}_{r=1}^k$ may be chosen to satisfy the differential equations as follows:

$$
\begin{aligned}
\frac{\mathrm{d}}{\mathrm{d}\varepsilon}\mathscr{E}_1(\varepsilon)&=A^T(\varepsilon)\mathscr{H}_1(\varepsilon)+\mathscr{H}_1(\varepsilon)A(\varepsilon)+C^T(\varepsilon)Q(\varepsilon)C(\varepsilon)\\
&-\mathscr{H}_1(\varepsilon)B(\varepsilon)R^{-1}(\varepsilon)B^T(\varepsilon)\sum_{s=1}^{k}\hat{\mu}_s\mathscr{H}_s(\varepsilon)\\
&-\sum_{r=1}^{k}\hat{\mu}_r\mathscr{H}_r(\varepsilon)B(\varepsilon)R^{-1}(\varepsilon)B^T(\varepsilon)\mathscr{H}_1(\varepsilon)\\
&+\sum_{r=1}^{k}\hat{\mu}_r\mathscr{H}_r(\varepsilon)B(\varepsilon)R^{-1}(\varepsilon)B^T(\varepsilon)\sum_{s=1}^{k}\hat{\mu}_s\mathscr{H}_s(\varepsilon), \qquad (2.40)
\end{aligned}
$$

$$
\begin{aligned}
\frac{\mathrm{d}}{\mathrm{d}\varepsilon}\mathscr{E}_r(\varepsilon)&=A^T(\varepsilon)\mathscr{H}_r(\varepsilon)+\mathscr{H}_r(\varepsilon)A(\varepsilon)\\
&-\mathscr{H}_r(\varepsilon)B(\varepsilon)R^{-1}(\varepsilon)B^T(\varepsilon)\sum_{s=1}^{k}\hat{\mu}_s\mathscr{H}_s(\varepsilon)\\
&-\sum_{s=1}^{k}\hat{\mu}_s\mathscr{H}_s(\varepsilon)B(\varepsilon)R^{-1}(\varepsilon)B^T(\varepsilon)\mathscr{H}_r(\varepsilon)\\
&+\sum_{s=1}^{r-1}\frac{2r!}{s!(r-s)!}\mathscr{H}_s(\varepsilon)G(\varepsilon)WG^T(\varepsilon)\mathscr{H}_{r-s}(\varepsilon), \quad 2\le r\le k, \qquad (2.41)
\end{aligned}
$$

$$\frac{d}{d\varepsilon}\breve{\mathscr{T}}_1(\varepsilon) = A^T(\varepsilon)\breve{\mathscr{D}}_1(\varepsilon) - C^T(\varepsilon)Q(\varepsilon)\zeta(\varepsilon)$$
$$- \sum_{s=1}^{k} \hat{\mu}_s \mathscr{H}_s(\varepsilon)B(\varepsilon)R^{-1}(\varepsilon)B^T(\varepsilon)\breve{\mathscr{D}}_1(\varepsilon)$$
$$- \mathscr{H}_1(\varepsilon)B(\varepsilon)R^{-1}(\varepsilon)B^T(\varepsilon)\sum_{s=1}^{k} \hat{\mu}_s \breve{\mathscr{D}}_s(\varepsilon)$$
$$+ \sum_{s=1}^{k} \hat{\mu}_s \mathscr{H}_s(\varepsilon)B(\varepsilon)R^{-1}(\varepsilon)B^T(\varepsilon)\sum_{s=1}^{k} \hat{\mu}_s \breve{\mathscr{D}}_s(\varepsilon), \tag{2.42}$$

$$\frac{d}{d\varepsilon}\breve{\mathscr{T}}_r(\varepsilon) = A^T(\varepsilon)\breve{\mathscr{D}}_r(\varepsilon) - \sum_{s=1}^{k} \hat{\mu}_s \mathscr{H}_s(\varepsilon)B(\varepsilon)R^{-1}(\varepsilon)B^T(\varepsilon)\breve{\mathscr{D}}_r(\varepsilon)$$
$$- \mathscr{H}_r(\varepsilon)B(\varepsilon)R^{-1}(\varepsilon)B^T(\varepsilon)\sum_{s=1}^{k} \hat{\mu}_s \breve{\mathscr{D}}_s(\varepsilon), \quad 2 \le r \le k, \tag{2.43}$$

$$\frac{d}{d\varepsilon}\mathscr{T}_1(\varepsilon) = \mathrm{Tr}\left\{\mathscr{H}_1(\varepsilon)G(\varepsilon)WG^T(\varepsilon)\right\} - 2\breve{\mathscr{D}}_1^T(\varepsilon)B(\varepsilon)R^{-1}(\varepsilon)B^T(\varepsilon)\sum_{s=1}^{k} \hat{\mu}_s \breve{\mathscr{D}}_s(\varepsilon)$$
$$+ \zeta^T(\varepsilon)Q(\varepsilon)\zeta(\varepsilon) + \sum_{s=1}^{k} \hat{\mu}_s \breve{\mathscr{D}}_s^T(\varepsilon)B(\varepsilon)R^{-1}(\varepsilon)B^T(\varepsilon)\sum_{s=1}^{k} \hat{\mu}_s \breve{\mathscr{D}}_s(\varepsilon), \tag{2.44}$$

$$\frac{d}{d\varepsilon}\mathscr{T}_r(\varepsilon) = \mathrm{Tr}\left\{\mathscr{H}_r(\varepsilon)G(\varepsilon)WG^T(\varepsilon)\right\}$$
$$- 2\breve{\mathscr{D}}_r^T(\varepsilon)B(\varepsilon)R^{-1}(\varepsilon)B^T(\varepsilon)\sum_{s=1}^{k} \hat{\mu}_s \breve{\mathscr{D}}_s(\varepsilon), \quad 2 \le r \le k. \tag{2.45}$$

The affine input and feedback gain specified in Eqs. (2.37) and (2.38) which are now applied along the solution trajectories of the Eqs. (2.27)–(2.29) yield the following results

$$\frac{d}{d\varepsilon}\mathscr{H}_1(\varepsilon) = -A^T(\varepsilon)\mathscr{H}_1(\varepsilon) - \mathscr{H}_1(\varepsilon)A(\varepsilon) + \mathscr{H}_1(\varepsilon)B(\varepsilon)R^{-1}(\varepsilon)B^T(\varepsilon)\sum_{s=1}^{k} \hat{\mu}_s \mathscr{H}_s(\varepsilon)$$
$$+ \sum_{s=1}^{k} \hat{\mu}_s \mathscr{H}_s(\varepsilon)B(\varepsilon)R^{-1}(\varepsilon)B^T(\varepsilon)\mathscr{H}_1(\varepsilon) - C^T(\varepsilon)Q(\varepsilon)C(\varepsilon)$$
$$- \sum_{s=1}^{k} \hat{\mu}_s \mathscr{H}_s(\varepsilon)B(\varepsilon)R^{-1}(\varepsilon)B^T(\varepsilon)\sum_{s=1}^{k} \hat{\mu}_s \mathscr{H}_s(\varepsilon), \tag{2.46}$$

$$\frac{d}{d\varepsilon}\mathscr{H}_r(\varepsilon) = -A^T(\varepsilon)\mathscr{H}_r(\varepsilon) - \mathscr{H}_r(\varepsilon)A(\varepsilon) + \mathscr{H}_r(\varepsilon)B(\varepsilon)R^{-1}(\varepsilon)B^T(\varepsilon)\sum_{s=1}^{k}\hat{\mu}_s\mathscr{H}_s(\varepsilon)$$
$$+\sum_{s=1}^{k}\hat{\mu}_s\mathscr{H}_s(\varepsilon)B(\varepsilon)R^{-1}(\varepsilon)B^T(\varepsilon)\mathscr{H}_r(\varepsilon)$$
$$-\sum_{s=1}^{r-1}\frac{2r!}{s!(r-s)!}\mathscr{H}_s(\varepsilon)G(\varepsilon)WG^T(\varepsilon)\mathscr{H}_{r-s}(\varepsilon), \quad 2 \le r \le k, \tag{2.47}$$

$$\frac{d}{d\varepsilon}\breve{\mathscr{D}}_1(\varepsilon) = -A^T(\varepsilon)\breve{\mathscr{D}}_1(\varepsilon) + \sum_{s=1}^{k}\hat{\mu}_s\mathscr{H}_s(\varepsilon)B(\varepsilon)R^{-1}(\varepsilon)B^T(\varepsilon)\breve{\mathscr{D}}_1(\varepsilon)$$
$$+C^T(\varepsilon)Q(\varepsilon)\zeta(\varepsilon) + \mathscr{H}_1(\varepsilon)B(\varepsilon)R^{-1}(\varepsilon)B^T(\varepsilon)\sum_{s=1}^{k}\hat{\mu}_s\breve{\mathscr{D}}_s(\varepsilon)$$
$$-\sum_{s=1}^{k}\hat{\mu}_s\mathscr{H}_s(\varepsilon)B(\varepsilon)R^{-1}(\varepsilon)B^T(\varepsilon)\sum_{s=1}^{k}\hat{\mu}_s\breve{\mathscr{D}}_s(\varepsilon), \tag{2.48}$$

$$\frac{d}{d\varepsilon}\breve{\mathscr{D}}_r(\varepsilon) = -A^T(\varepsilon)\breve{\mathscr{D}}_r(\varepsilon) + \mathscr{H}_r(\varepsilon)B(\varepsilon)R^{-1}(\varepsilon)B^T(\varepsilon)\sum_{s=1}^{k}\hat{\mu}_s\breve{\mathscr{D}}_s(\varepsilon)$$
$$+\sum_{s=1}^{k}\hat{\mu}_s\mathscr{H}_s(\varepsilon)B(\varepsilon)R^{-1}(\varepsilon)B^T(\varepsilon)\breve{\mathscr{D}}_r(\varepsilon), \quad 2 \le r \le k, \tag{2.49}$$

$$\frac{d}{d\varepsilon}\mathscr{D}_1(\varepsilon) = -\mathrm{Tr}\left\{\mathscr{H}_1(\varepsilon)G(\varepsilon)WG^T(\varepsilon)\right\} + 2\breve{\mathscr{D}}_1^T(\varepsilon)B(\varepsilon)R^{-1}(\varepsilon)B^T(\varepsilon)\sum_{s=1}^{k}\hat{\mu}_s\breve{\mathscr{D}}_s(\varepsilon)$$
$$-\sum_{s=1}^{k}\hat{\mu}_s\breve{\mathscr{D}}_s^T(\varepsilon)B(\varepsilon)R^{-1}(\varepsilon)B^T(\varepsilon)\sum_{s=1}^{k}\hat{\mu}_s\breve{\mathscr{D}}_s(\varepsilon) - \zeta^T(\varepsilon)Q(\varepsilon)\zeta(\varepsilon), \tag{2.50}$$

$$\frac{d}{d\varepsilon}\mathscr{D}_r(\varepsilon) = -\mathrm{Tr}\left\{\mathscr{H}_r(\varepsilon)G(\varepsilon)WG^T(\varepsilon)\right\}$$
$$+2\breve{\mathscr{D}}_r^T(\varepsilon)B(\varepsilon)R^{-1}(\varepsilon)B^T(\varepsilon)\sum_{s=1}^{k}\hat{\mu}_s\breve{\mathscr{D}}_s(\varepsilon), \quad 2 \le r \le k. \tag{2.51}$$

whereby the terminal-value conditions $\mathscr{H}_1(t_f) = C^T(t_f)Q_fC(t_f)$, $\mathscr{H}_r(t_f) = 0$ for $2 \le r \le k$; $\breve{\mathscr{D}}_1(t_f) = -C^T(t_f)Q_f\zeta(t_f)$, $\breve{\mathscr{D}}_r(t_f) = 0$ for $2 \le r \le k$ and $\mathscr{D}_1(t_f) = \zeta^T(t_f)Q_f\zeta(t_f)$, $\mathscr{D}_r(t_f) = 0$ for $2 \le r \le k$.

Finally, the boundary condition of $\mathscr{W}(\varepsilon, \mathscr{Y}, \breve{\mathscr{Z}}, \mathscr{Z})$ implies that

$$x_0^T \sum_{r=1}^{k} \mu_r (\mathscr{H}_r(t_0) + \mathscr{E}_r(t_0)) x_0 + 2x_0^T \sum_{r=1}^{k} \mu_r (\breve{\mathscr{D}}_r(t_0) + \breve{\mathscr{T}}_r(t_0))$$
$$+ \sum_{r=1}^{k} \mu_r (\mathscr{D}_r(t_0) + \mathscr{T}_r(t_0)) = x_0^T \sum_{r=1}^{k} \mu_r \mathscr{H}_r(t_0) x_0 + 2x_0^T \sum_{r=1}^{k} \mu_r \breve{\mathscr{D}}_r(t_0) + \sum_{r=1}^{k} \mu_r \mathscr{D}_r(t_0).$$

The initial conditions for the forward-in-time differential equations (2.40)–(2.45) are therefore determined as follows: $\mathscr{E}_r(t_0) = 0$, $\breve{\mathscr{T}}_r(t_0) = 0$ and $\mathscr{T}_r(t_0) = 0$. Further on, the optimal affine input (2.37), and state-feedback feedback gain (2.38) minimizing the performance index (2.26) become

$$l^*(\varepsilon) = -R^{-1}(\varepsilon) B^T(\varepsilon) \sum_{r=1}^{k} \hat{\mu}_r \breve{\mathscr{D}}_r^*(\varepsilon),$$

$$K^*(\varepsilon) = -R^{-1}(\varepsilon) B^T(\varepsilon) \sum_{r=1}^{k} \hat{\mu}_r \mathscr{H}_r^*(\varepsilon).$$

The theorem that follows contains a controller design procedure which is able to track a prescribed function of time in accordance with the performance appraisal consisted of performance values and risks. The statistical optimal tracking controller requires a state-feedback control gain and an affine input that results from the backward-in-time solutions of the coupled differential equations.

Theorem 2.5.2 (Statistical Optimal Control Solution). *Consider the linear-quadratic class of tracking problems described by Eqs. (2.1)–(2.2), wherein the process noise $w(t) \in \mathbb{R}^p$ is the p-dimensional Wiener process starting from t_0, independent of the initial system state x_0 and defined on a complete probability space $(\Omega, \mathscr{F}, \mathscr{P})$ over $[t_0, t_f]$ with the correlation of increments $E\{[w(\tau_1) - w(\tau_2)][w(\tau_1) - w(\tau_2)]^T\} = W|\tau_1 - \tau_2|$ and $W > 0$. The admissible control $u \in L^2_{\mathscr{F}_t}(\Omega; \mathscr{C}([t_0, t_f]; \mathbb{R}^m))$ is selected so that the resulting output $y \in L^2_{\mathscr{F}_t}(\Omega; \mathscr{C}([t_0, t_f]; \mathbb{R}^r))$ best approximates a priori trajectories $\zeta \in L^2(\mathscr{C}([t_0, t_f]; \mathbb{R}^r))$ in the sense of Eq. (2.26) in which the terminal penalty error weighting $Q_f \in \mathbb{R}^{r \times r}$, the error weighting $Q \in \mathscr{C}([t_0, t_f]; \mathbb{R}^{r \times r})$, and the control input weighting $R \in \mathscr{C}([t_0, t_f]; \mathbb{R}^{m \times m})$ are symmetric and positive semidefinite with $R(t)$ invertible.*

Assume $k \in \mathbb{Z}^+$ and the sequence $\mu = \{\mu_r \geq 0\}_{r=1}^k$ with $\mu_1 > 0$ are fixed. Then, the statistical optimal control solution based on the state feedback measurements for the finite-horizon tracking problem is implemented by

$$u^*(t) = K^*(t) x^*(t) + l^*(t), \quad t = t_0 + t_f - \tau, \forall \tau \in [t_0, t_f], \tag{2.52}$$

$$K^*(\tau) = -R^{-1}(\tau) B^T(\tau) \sum_{r=1}^{k} \hat{\mu}_r \mathscr{H}_r^*(\tau), \tag{2.53}$$

$$l^*(\tau) = -R^{-1}(\tau)B^T(\tau)\sum_{r=1}^{k}\hat{\mu}_r\breve{\mathscr{D}}_r^*(\tau), \tag{2.54}$$

whereby $\hat{\mu}_r \triangleq \mu_r/\mu_1$ *and whenever* $\{\mathscr{H}_r^*(\tau)\}_{r=1}^k$ *and* $\{\breve{\mathscr{D}}_r^*(\tau)\}_{r=1}^k$ *are the solutions of the backward-in-time matrix-valued differential equations*

$$\frac{d}{d\tau}\mathscr{H}_1^*(\tau) = -\left[A(\tau)+B(\tau)K^*(\tau)\right]^T\mathscr{H}_1^*(\tau) - \mathscr{H}_1^*(\tau)\left[A(\tau)+B(\tau)K^*(\tau)\right]$$
$$-C^T(\tau)Q(\tau)C(\tau) - K^{*T}(\tau)R(\tau)K^*(\tau), \tag{2.55}$$
$$\frac{d}{d\tau}\mathscr{H}_r^*(\tau) = -\left[A(\tau)+B(\tau)K^*(\tau)\right]^T\mathscr{H}_r^*(\tau) - \mathscr{H}_r^*(\tau)\left[A(\tau)+B(\tau)K^*(\tau)\right]$$
$$-\sum_{s=1}^{r-1}\frac{2r!}{s!(r-s)!}\mathscr{H}_s^*(\tau)G(\tau)WG^T(\tau)\mathscr{H}_{r-s}^*(\tau), \quad 2\le r\le k \tag{2.56}$$

and the backward-in-time vector-valued differential equations

$$\frac{d}{d\tau}\breve{\mathscr{D}}_1^*(\tau) = -\left[A(\tau)+B(\tau)K^*(\tau)\right]^T\breve{\mathscr{D}}_1^*(\tau) - \mathscr{H}_1(\tau)B(\tau)l^*(\tau)$$
$$-K^{*T}(\tau)R(\tau)l^*(\tau) + C^T(\tau)Q(\tau)\zeta(\tau), \tag{2.57}$$
$$\frac{d}{d\tau}\breve{\mathscr{D}}_r^*(\tau) = -\left[A(\tau)+B(\tau)K^*(\tau)\right]^T\breve{\mathscr{D}}_r^*(\tau) - \mathscr{H}_r(\tau)B(\tau)l^*(\tau), 2\le r\le k \tag{2.58}$$

provided that the terminal-value conditions $\mathscr{H}_1^*(t_f) = C^T(t_f)Q_fC(t_f)$, $\mathscr{H}_r^*(t_f) = 0$ *for* $2\le r\le k$ *and* $\breve{\mathscr{D}}_1^*(t_f) = -C^T(t_f)Q_f\zeta(t_f)$, $\breve{\mathscr{D}}_r^*(t_f) = 0$ *for* $2\le r\le k$.

2.6 Chapter Summary

In this chapter, an optimal control problem for a wide class of tracking systems is formulated in which the objective is minimization of a finite, linear combination of cumulants of integral quadratic cost over linear, memoryless, full-state-feedback control laws. The standard linear tracking system constraint on a finite time interval with additive Wiener noise and a non-random initial state underlies the problem formulation. Because of the linearity assumptions in the problem statement, it can be formulated as a non-stochastic optimization problem utilizing equations for cost cumulants developed in this exposition. Furthermore, since this problem formulation is parameterized both by the number of cumulants and by the scalar coefficients in the linear combination, it defines a very general LQG and risk sensitive problem classes. For instance, the special case where only the first cost cumulant is minimized is, of course, the well-known minimum mean LQG problem and whereas a denumerable linear combination of cost cumulants is minimized is the continued risk-sensitive control objective. It should also be noted that, although the optimization criterion of the statistical optimal control problem represents a competition among cumulant values, the ultimate objective herein is to introduce parametric designs of freedom in the class of feedback control laws which will result from the problem solution. These parametric designs of freedom have been exploited to achieve desirable closed-loop system properties. Finally, the general solution of the statistical optimal control problem for the class of linear-quadratic tracking systems is presented and is determined by a feedback statistical optimal control obtained by a set of coupled Riccati-type differential equations and time-dependent tracking variables found by solving an auxiliary set of coupled differential equations (incorporating the desired trajectory) backward from a stable final time. The issue of existence of solution to the optimization problem becomes that of existence of solutions to the Riccati-type equations. Conditions ensuring existence of solutions to these equations are being worked out. In fact, for values of linear combination coefficients outside certain finite ranges, the equations exhibit finite escape time behavior. On the other hand, for limited ranges of the combination coefficient values, the equations are well behaved and yield steady-state solutions as evident in numerous controller designs for civil structural control and aerospace applications.

References

1. Pham, K.D.: Minimax design of statistics-based control with noise uncertainty for highway bridges. In: Proceedings of DETC 2005/2005 ASME 20th Biennial Conference on Mechanical Vibration and Noise, DETC2005-84593 (2005)
2. Pham, K.D., Sain, M.K., Liberty, S.R.: Statistical control for smart base-isolated buildings via cost cumulants and output feedback paradigm. In: Proceedings of American Control Conference, pp. 3090–3095 (2005)
3. Pham, K.D., Jin, G., Sain, M.K., Spencer, B.F. Jr., Liberty, S.R.: Generalized LQG techniques for the wind benchmark problem. Special Issue of ASCE J. Eng. Mech. Struct. Contr. Benchmark Problem **130**(4), 466–470 (2004)
4. Pham, K.D., Sain, M.K., Liberty, S.R.: Infinite horizon robustly stable seismic protection of cable-stayed bridges using cost cumulants. In: Proceedings of American Control Conference, pp. 691–696 (2004)
5. Pham, K.D., Sain, M.K., Liberty, S.R.: Cost cumulant control: state-feedback, finite-horizon paradigm with application to seismic protection. In: Miele, A. (ed.) Special Issue of Journal of Optimization Theory and Applications, vol. 115(3), pp. 685–710. Kluwer Academic/Plenum Publishers, New York (2002)
6. Pham, K.D., Sain, M.K., Liberty, S.R.: Robust cost-cumulants based algorithm for second and third generation structural control benchmarks. In: Proceedings of American Control Conference, pp. 3070–3075 (2002)
7. Mou, L., Liberty, S.R., Pham, K.D., Sain, M.K.: Linear cumulant control and its relationship to risk-sensitive control. In: Proceedings of 38th Annual Allerton Conference on Communication, Control, and Computing, pp. 422–430 (2000)
8. Jin, G., Pham, K.D., Spencer, B.F. Jr., Sain, M.K., Liberty, S.R.: Protecting tall buildings under stochastic winds using multiple cost cumulants. In: Proceedings of the 8th ASCE Specialty Conference on Probabilistic Mechanics and Structural Reliability Conference (2000)

9. Pham, K.D., Sain, M.K., Liberty, S.R., Spencer, B.F. Jr.: Optimum multiple cost cumulants for protection of civil structures. In: Proceedings of the 8th ASCE Specialty Conference on Probabilistic Mechanics and Structural Reliability Conference (2000)
10. Jin, G., Pham, K.D., Spencer, B.F. Jr., Sain, M.K., Liberty, S.R.: A study of the ASCE wind benchmark problem by generalized LQG techniques. In: Proceedings of the 2nd European Conference on Structural Control (2000)
11. Pham, K.D., Liberty, S.R., Sain, M.K., Spencer, B.F. Jr. First generation seismic-AMD benchmark: Robust structural protection by the cost cumulant control paradigm. In: Proceedings of American Control Conference, pp. 1–5 (2000)
12. Pham, K.D., Jin, G., Sain, M.K., Spencer, B.F. Jr., Liberty, S.R.: Third generation wind-AMD benchmark: Cost cumulant control methodology for wind excited tall buildings. In: Proceedings of the 14th ASCE Engineering Mechanics Conference (2000)
13. Pham, K.D., Sain, M.K., Liberty, S.R., Spencer, B.F. Jr.: The role and use of optimal cost cumulants for protection of civil structures. In: Proceedings of the 14th ASCE Engineering Mechanics Conference (2000)
14. Pham, K.D., Liberty, S.R., Sain, M.K.: Evaluating cumulant controllers on a benchmark structure protection problem in the presence of classic earthquakes. In: Proceedings of 37th Annual Allerton Conference on Communication, Control, and Computing, pp. 617–626 (1999)
15. Pham, K.D., Liberty, S.R., Sain, M.K., Spencer, B.F. Jr.: Generalized risk sensitive building control: Protecting civil structures with multiple cost cumulants. In: Proceedings of American Control Conference, pp. 500–504 (1999)
16. Pham, K.D., Liberty, S.R., Sain, M.K.: Linear optimal cost cumulant control: A k-cumulant problem class. In: Proceedings of 36th Annual Allerton Conference on Communication, Control, and Computing, pp. 460–469 (1998)
17. Luenberger, D.G.: Tracking of goal seeking vehicles. IEEE Trans. Automat. Contr. **13**(2), 74–77 (1968)
18. Fleming, W.H., Rishel, R.W.: Deterministic and Stochastic Optimal Control. Springer, New York (1975)
19. Pham, K.D.: New risk-averse control paradigm for stochastic two-time-scale systems and performance robustness. In: Miele, A. (ed.) J. Optim. Theor. Appl. **146**(2), 511–537 (2010)
20. Pham, K.D.: Performance-reliability aided decision making in multiperson quadratic decision games against jamming and estimation confrontations. In: Giannessi, F. (ed.) J. Optim. Theor. Appl. **149**(3), 559–629 (2011)

Chapter 3
Overtaking Tracking Problems in Risk-Averse Control

Abstract Among the important results herein is the performance information analysis of forecasting higher-order characteristics of a general criterion of performance associated with a stochastic tracking system which is closely supervised by a reference command input and a desired trajectory. Both compactness from logic of state-space model description and quantitativity from probabilistic knowledge of stochastic disturbances are exploited to therefore allow accurate prediction of the effects of chi-squared randomness on performance distribution of the optimal tracking problem. Information about performance-measure statistics is further utilized in the synthesis of statistical optimal controllers which are thus capable of shaping the distribution of tracking performance without reliance on computationally intensive Monte Carlo analysis as needed in post-design performance assessment. As a by-product, the recent results can potentially be applicable to another substantially larger class of optimal tracking systems whereby local representations with only first two statistics for non-Gaussian random distributions of exogenous disturbances and uncertain environments may be sufficient.

3.1 Introduction

Thus far statistical optimal control is considered only with linear-quadratic tracking problems. Obviously, many interesting optimization problems involve closely guided tracking criteria. For instance, a class of overtaking tracking problems is central to the study of physical systems as it is to the synthesis of feedback systems that are able to track a-priori scheduling signals and target control references. Interested readers may consult [1–3] to appreciate the scope of the concepts involved in designing feedback controls for deterministic systems that optimize quadratic performance indices of reference signals. The motivation in writing the present chapter is to use performance information to affect achievable performance in risk-averse decision making and feedback design. The recent work proposed by the author has begun to address some key and unique aspects as follows. First, there is a

K.D. Pham, *Linear-Quadratic Controls in Risk-Averse Decision Making*,
SpringerBriefs in Optimization, DOI 10.1007/978-1-4614-5079-5_3,

recognition process that comprehends the significance of linear-quadratic structure of the stochastic tracking dynamics and incorporates this special property in the criterion of performance. Hence, the measure of performance is, in fact, a random variable with chi-squared type and thus, all random sample-path realizations from the underlying stochastic process will lead to riskier and uncertain performance. The second aspect involves the linkage of a priori knowledge of probabilistic distribution of the underlying stochastic process with system performance distribution and thus describes how higher-order statistics associated with the performance-measure are exploited to project future status of performance uncertainty. The third aspect, which is distinct from the traditional average performance optimization, is a general measure of performance riskiness as being a finite linear combination of performance-measure statistics of choice that the feedback controller uses for its adaptive control decisions. Since the account [4] has initially dealt with the issue of performance robustness in stochastic tracking problems, it is therefore natural to further extend the existing tracking results with additional command input references.

Notional advantages offered by the proposed paradigm are especially effective for uncertainty analysis. That is, qualitative assessment of the impact of uncertainty caused by stochastic disturbances on system performance has long been recognized as an important and indispensable consideration in reliability-based design [5] and [6]. The research investigation is organized as follows. In Sect. 3.2 the tracking system description together with the definition of performance-measure statistics and their supporting equations associated with the chi-squared random measure of performance is presented. Problem statements for the resulting Mayer problem in dynamic programming are given in Sect. 3.3. Construction of a candidate function for the value function and the calculation of optimal feedback control accounting for multiple internalized goals of performance robustness are included in Sects. 3.4 and 3.5, while conclusions are drawn in Sect. 3.6.

3.2 Problem Description

Consider a general class of stochastic tracking systems, modeled on a finite horizon $[t_0, t_f]$ and governed by

$$dx(t) = (A(t)x(t) + B(t)u(t))dt + G(t)dw(t), \quad x(t_0) = x_0, \tag{3.1}$$

whereby the continuous-time coefficients $A \in \mathscr{C}([t_0,t_f];\mathbb{R}^{n\times n})$, $B \in \mathscr{C}([t_0,t_f];\mathbb{R}^{n\times m})$ and $G \in \mathscr{C}([t_0,t_f];\mathbb{R}^{n\times p})$ are deterministic, bounded matrix-valued functions. Uncertain environments and exogenous disturbances, $w(t) \in \mathbb{R}^p$ are characterized by an p-dimensional stationary Wiener process starting from t_0, independent of the known initial condition x_0, and defined with $\{\mathscr{F}_t\}_{t\geq t_0>0}$ being its filtration on a complete filtered probability space $(\Omega, \mathscr{F}, \{\mathscr{F}_t\}_{t\geq t_0>0}, \mathscr{P})$ over $[t_0,t_f]$ with the correlation of independent increments $E\{[w(\tau_1)-w(\tau_2)][w(\tau_1)-w(\tau_2)]^T\}=W|\tau_1-\tau_2|$ for all $\tau_1,\tau_2\in[t_0,t_f]$ and $W>0$. The set of admissible controls $L^2_{\mathscr{F}_t}(\Omega;\mathscr{C}([t_0,t_f];\mathbb{R}^m))$ belongs to the Hilbert space of $\mathbb{R}^m$-valued square-integrable processes on $[t_0,t_f]$ that are adapted to the σ-field $\mathscr{F}_t$ generated by $w(t)$ with $E\{\int_{t_0}^{t_f}u^T(\tau)u(\tau)d\tau\}<\infty$.

Associated with the admissible 2-tuple $(x(\cdot);u(\cdot))\in L^2_{\mathscr{F}_t}(\Omega;\mathscr{C}([t_0,t_f];\mathbb{R}^n))\times L^2_{\mathscr{F}_t}(\Omega;\mathscr{C}([t_0,t_f];\mathbb{R}^m))$ is a finite-horizon integral-quadratic performance measure $J:L^2_{\mathscr{F}_t}(\Omega;\mathscr{C}([t_0,t_f];\mathbb{R}^m))\mapsto\mathbb{R}^+$

$$J(u(\cdot))=(x(t_f)-\gamma(t_f))^TQ_f(x(t_f)-\gamma(t_f))+\int_{t_0}^{t_f}[(x(\tau)-\gamma(\tau))^TQ(\tau)(x(\tau)-\gamma(\tau)) + (u(\tau)-u_r(\tau))^TR(\tau)(u(\tau)-u_r(\tau))]d\tau, \tag{3.2}$$

whereby the desired trajectory $\gamma(\cdot)$ and reference control input $u_r(\cdot)$ are given, deterministic, bounded and piecewise-continuous functions on $[t_0,t_f]$. Design parameters $Q_f\in\mathbb{R}^{n\times n}$, $Q\in\mathscr{C}([t_0,t_f];\mathbb{R}^{n\times n})$, and invertible $R\in\mathscr{C}([t_0,t_f];\mathbb{R}^{m\times m})$ are deterministic, bounded, matrix-valued and positive semidefinite relative weightings of the terminal state, state trajectory, and control input.

Furthermore, as shown in [4], under linear, state-feedback control together with the fact of the linear-quadratic system, all higher-order statistics of the integral-quadratic performance measure have the quadratic-affine functional form. This common form of these higher-order statistics facilitates the definition of a risk-value aware performance index and the associated optimization problem. Therefore, the control law considered here is a linear time-varying feedback law generated from the tracking state $x(t)$ and reference control input $u_r(t)$ by

$$u(t)=K(t)x(t)+l_f(t)+u_r(t), \tag{3.3}$$

where both admissible vector-valued affine input $l_f\in\mathscr{C}([t_0,t_f];\mathbb{R}^m)$ and matrix-valued feedback gain $K\in\mathscr{C}([t_0,t_f];\mathbb{R}^{m\times n})$ are yet to be determined. Henceforth, for the admissible initial condition $(t_0,x_0)\in[t_0,t_f]\times\mathbb{R}^n$ and the feasible control policy (3.3), the dynamics of the generalized tracking problem are governed by the stochastic differential equation with the initial condition $x(t_0)=x_0$ known

$$dx(t)=[(A(t)+B(t)K(t))x(t)+B(t)(l_f(t)+u_r(t))]dt+G(t)dw(t) \tag{3.4}$$

and subject to the realized performance measure

$$J(K(\cdot),l_f(\cdot))=(x(t_f)-\gamma(t_f))^TQ_f(x(t_f)-\gamma(t_f))+\int_{t_0}^{t_f}[(x(\tau)-\gamma(\tau))^TQ(\tau)(x(\tau) -\gamma(\tau))+(K(\tau)x(\tau)+l_f(\tau))^TR(\tau)(K(\tau)x(\tau)+l_f(\tau))]d\tau. \tag{3.5}$$

3.3 A Framework for Performance-Measure Statistics

Up to now, the performance measure (3.5) is regarded as a random variable with the chi-squared type from which the uncertainty of performance distribution must be addressed via a complete set of higher-order statistics beyond the statistical averaging. Thus, the concepts and methods of characterization and management of higher-order statistics associated with the performance measure (3.5) are directly applicable to risk-averse control problems with performance risk consequences. Such performance-measure statistics are sometimes called cost cumulants for short and thus are utilized in the synthesis of state-feedback controllers to directly target the uncertainty of tracking performance.

In the development to follow, the initial condition (t_0, x_0) is treated as any arbitrary and yet admissible pair (τ, x_τ). Therefore, for the admissible affine input l_f and feedback gain K, the performance measure (3.5) is considered as the "performance-to-come," $J(\tau, x_\tau)$

$$J(\tau,x_\tau) = (x(t_f)-\gamma(t_f))^T Q_f (x(t_f)-\gamma(t_f)) + \int_\tau^{t_f} [(x(\tau)-\gamma(\tau))^T Q(\tau) \times (x(\tau)-\gamma(\tau)) + (K(\tau)x(\tau)+l_f(\tau))^T R(\tau)(K(\tau)x(\tau)+l_f(\tau))]\,\mathrm{d}\tau. \quad (3.6)$$

The moment-generating function of the "performance-to-come" (3.6) is defined by

$$\varphi(\tau,x_\tau;\theta) \triangleq E\{\exp(\theta J(\tau,x_\tau))\} \quad (3.7)$$

for all small parameters θ in an open interval about 0. Thus, the cumulant-generating function immediately follows

$$\psi(\tau,x_\tau;\theta) \triangleq \ln\{\varphi(\tau,x_\tau;\theta)\} \quad (3.8)$$

for all θ in some (possibly smaller) open interval about 0 while $\ln\{\cdot\}$ denotes the natural logarithmic transformation.

Theorem 3.3.1 (Cumulant-Generating Function). *Suppose that $\tau \in [t_0, t_f]$ is some running variable and θ is a small positive parameter. When $\varphi(\tau,x_\tau;\theta) \triangleq \rho(\tau,\theta)\exp\{x_\tau^T \Upsilon(\tau,\theta)x_\tau + 2x_\tau^T \eta(\tau,\theta)\}$ and $\upsilon(\tau,\theta) \triangleq \ln\{\rho(\tau,\theta)\}$, the cumulant-generating function that contains all the higher-order characteristics of the performance distribution, is then given by the expression*

$$\psi(\tau,x_\tau;\theta) = x_\tau^T \Upsilon(\tau,\theta)x_\tau + 2x_\tau^T \eta(\tau,\theta) + \upsilon(\tau,\theta), \quad (3.9)$$

where the cumulant-supporting variables $\Upsilon(\tau,\theta)$, $\eta(\tau,\theta)$, and $\upsilon(\tau,\theta)$ solve the time-backward differential equations:

$$\frac{\mathrm{d}}{\mathrm{d}\tau}\Upsilon(\tau,\theta) = -[A(\tau)+B(\tau)K(\tau)]^T \Upsilon(\tau,\theta) - \Upsilon(\tau,\theta)[A(\tau)+B(\tau)K(\tau)] - 2\Upsilon(\tau,\theta)G(\tau)WG^T(\tau)\Upsilon(\tau,\theta) - \theta\left[Q(\tau)+K^T(\tau)R(\tau)K(\tau)\right], \quad (3.10)$$

$$\frac{d}{d\tau}\eta(\tau,\theta) = -[A(\tau)+B(\tau)K(\tau)]^T\eta(\tau,\theta) - \Upsilon(\tau,\theta)B(\tau)\left[l_f(\tau)+u_r(\tau)\right] - \theta\left[K^T(\tau)R(\tau)l_f(\tau) - Q(\tau)\gamma(\tau)\right], \quad (3.11)$$

$$\frac{d}{d\tau}\upsilon(\tau,\theta) = -\mathrm{Tr}\left\{\Upsilon(\tau,\theta)G(\tau)WG^T(\tau)\right\} - 2\eta^T(\tau,\theta)B(\tau)\left[l_f(\tau)+u_r(\tau)\right] - \theta\left[l_f^T(\tau)R(\tau)l_f(\tau) + \gamma^T(\tau)Q(\tau)\gamma(\tau)\right] \quad (3.12)$$

with the terminal-value conditions $\Upsilon(t_f,\theta) = \theta Q_f$, $\eta(t_f,\theta) = -\theta Q_f\gamma(t_f)$, *and* $\upsilon(t_f,\theta) = \theta\gamma^T(t_f)Q_f\gamma(t_f)$.

Proof. For simplicity, it is convenient to have $\varpi(\tau,x_\tau;\theta) \triangleq \exp\{\theta J(\tau,x_\tau)\}$ and $\varphi(\tau,x_\tau;\theta) \triangleq E\{\varpi(\tau,x_\tau;\theta)\}$. In addition, it follows that

$$\frac{d}{d\tau}\varphi(\tau,x_\tau;\theta) = -\theta\Big\{x_\tau^T[Q(\tau)+K^T(\tau)R(\tau)K(\tau)]x_\tau + l_f^T(\tau)R(\tau)l_f(\tau) + 2x_\tau^T[K^T(\tau)R(\tau)l_f(\tau) - Q(\tau)\gamma(\tau)] + \gamma^T(\tau)Q(\tau)\gamma(\tau)\Big\}\varphi(\tau,x_\tau;\theta). \quad (3.13)$$

Using the standard Ito's formula, it yields

$$\begin{aligned} d\varphi(\tau,x_\tau;\theta) &= E\{d\varpi(\tau,x_\tau;\theta)\} \\ &= \varphi_\tau(\tau,x_\tau;\theta)\,d\tau + \varphi_{x_\tau}(\tau,x_\tau;\theta)\left[(A(\tau)+B(\tau)(K(\tau))x_\tau \right. \\ &\quad \left. + B(\tau)(l_f(\tau)+u_r(\tau))\right]d\tau + \frac{1}{2}\mathrm{Tr}\left\{\varphi_{x_\tau x_\tau}(\tau,x_\tau;\theta)\,G(\tau)WG^T(\tau)\right\}d\tau, \end{aligned}$$

which, under the definition $\varphi(\tau,x_\tau;\theta) \triangleq \rho(\tau,\theta)\exp\{x_\tau^T\Upsilon(\tau,\theta)x_\tau + 2x_\tau^T\eta(\tau,\theta)\}$ in addition with its partial derivatives, leads to the total derivative of $\varphi(\tau,x_\tau;\theta)$ with respect to time

$$\begin{aligned} &\frac{d}{d\tau}\varphi(\tau,x_\tau;\theta) \\ &= \left\{\frac{\frac{d}{d\tau}\rho(\tau,\theta)}{\rho(\tau,\theta)} + x_\tau^T\frac{d}{d\tau}\Upsilon(\tau,\theta)x_\tau + 2x_\tau^T\frac{d}{d\tau}\eta(\tau,\theta)\right. \\ &\quad + x_\tau^T[A(\tau)+B(\tau)K(\tau)]^T\Upsilon(\tau,\theta)x_\tau + x_\tau^T\Upsilon(\tau,\theta)[A(\tau)+B(\tau)K(\tau)]x_\tau \\ &\quad + 2x_\tau^T[A(\tau)+B(\tau)K(\tau)]^T\eta(\tau,\theta) + 2x_\tau^T\Upsilon(\tau,\theta)B(\tau)[l_f(\tau)+u_r(\tau)] \\ &\quad + 2\eta^T(\tau,\theta)B(\tau)[l_f(\tau)+u_r(\tau)] + \mathrm{Tr}\left\{\Upsilon(\tau,\theta)G(\tau)WG^T(\tau)\right\} \\ &\quad \left. + 2x_\tau^T\Upsilon(\tau,\theta)G(\tau)WG^T(\tau)\Upsilon(\tau,\theta)x_\tau\right\}\varphi(\tau,x_\tau;\theta). \end{aligned} \quad (3.14)$$

Substituting the result (3.13) into the left-hand side of Eq. (3.14) and having both linear and quadratic terms independent of x_τ require that

$$\frac{\mathrm{d}}{\mathrm{d}\tau}\Upsilon(\tau,\theta) = -[A(\tau)+B(\tau)K(\tau)]^T\Upsilon(\tau,\theta) - \Upsilon(\tau,\theta)[A(\tau)+B(\tau)K(\tau)]$$
$$-2\Upsilon(\tau,\theta)G(\tau)WG^T(\tau)\Upsilon(\tau,\theta) - \theta\left[Q(\tau)+K^T(\tau)R(\tau)K(\tau)\right],$$

$$\frac{\mathrm{d}}{\mathrm{d}\tau}\eta(\tau,\theta) = -[A(\tau)+B(\tau)K(\tau)]^T\eta(\tau,\theta) - \Upsilon(\tau,\theta)B(\tau)\left[l_f(\tau)+u_r(\tau)\right]$$
$$-\theta\left[K^T(\tau)R(\tau)l_f(\tau) - Q(\tau)\gamma(\tau)\right],$$

$$\frac{\mathrm{d}}{\mathrm{d}\tau}\upsilon(\tau,\theta) = -\mathrm{Tr}\left\{\Upsilon(\tau,\theta)G(\tau)WG^T(\tau)\right\} - 2\eta^T(\tau,\theta)B(\tau)\left[l_f(\tau)+u_r(\tau)\right]$$
$$-\theta\left[l_f^T(\tau)R(\tau)l_f(\tau)+\gamma^T(\tau)Q(\tau)\gamma(\tau)\right].$$

At the final time $\tau = t_f$, it follows that

$$\varphi(t_f,x(t_f);\theta) = \rho(t_f,\theta)\exp\left\{x^T(t_f)\Upsilon(t_f,\theta)x(t_f)+2x^T(t_f)\eta(t_f,\theta)\right\}$$
$$= E\left\{\exp\left\{\theta[x(t_f)-\gamma(t_f)]^T Q_f[x(t_f)-\gamma(t_f)]\right\}\right\},$$

which, in turn yields the terminal-value conditions as $\Upsilon(t_f,\theta) = \theta Q_f$, $\eta(t_f,\theta) = -\theta Q_f\gamma(t_f)$, $\rho(t_f,\theta) = \exp\left\{\theta\gamma^T(t_f)Q_f\gamma(t_f)\right\}$ and $\upsilon(t_f,\theta) = \theta\gamma^T(t_f)Q_f\gamma(t_f)$. □

Notice that the expression for cumulant-generating function (3.9) for the generalized performance measure (3.5) indicates that additional affine and trailing terms take into account of dynamics mismatched in the transient responses.

By definition, higher-order performance-measure statistics that encapsulate the uncertain nature of tracking performance can now be generated via a Maclaurin series expansion of the cumulant-generating function (3.9)

$$\psi(\tau,x_\tau;\theta) \triangleq \sum_{r=1}^{\infty}\kappa_r\frac{\theta^r}{r!} = \sum_{r=1}^{\infty}\left.\frac{\partial^{(r)}}{\partial\theta^{(r)}}\psi(\tau,x_\tau;\theta)\right|_{\theta=0}\frac{\theta^r}{r!} \tag{3.15}$$

from which κ_r is denoted as the rth-performance-measure statistics or the rth-cost cumulant. Moreover, the series expansion coefficients are thus obtained by using the cumulant-generating function (3.9)

$$\left.\frac{\partial^{(r)}}{\partial\theta^{(r)}}\psi(\tau,x_\tau;\theta)\right|_{\theta=0} = x_\tau^T\left.\frac{\partial^{(r)}}{\partial\theta^{(r)}}\Upsilon(\tau,\theta)\right|_{\theta=0} x_\tau + 2x_\tau^T\left.\frac{\partial^{(r)}}{\partial\theta^{(r)}}\eta(\tau,\theta)\right|_{\theta=0}$$
$$+\left.\frac{\partial^{(r)}}{\partial\theta^{(r)}}\upsilon(\tau,\theta)\right|_{\theta=0}. \tag{3.16}$$

In view of the results (3.15) and (3.16), the rth performance-measure statistic for the generalized tracking problem therefore follows

$$\kappa_r = x_\tau^T \frac{\partial^{(r)}}{\partial\theta^{(r)}}\Upsilon(\tau,\theta)\bigg|_{\theta=0} x_\tau + 2x_\tau^T \frac{\partial^{(r)}}{\partial\theta^{(r)}}\eta(\tau,\theta)\bigg|_{\theta=0} + \frac{\partial^{(r)}}{\partial\theta^{(r)}}\upsilon(\tau,\theta)\bigg|_{\theta=0} \tag{3.17}$$

for any finite $1 \leq r < \infty$.

For notational convenience, the following definitions

$$H_r(\tau) \triangleq \frac{\partial^{(r)}}{\partial\theta^{(r)}}\Upsilon(\tau,\theta)\bigg|_{\theta=0}; \breve{D}_r(\tau) \triangleq \frac{\partial^{(r)}}{\partial\theta^{(r)}}\eta(\tau,\theta)\bigg|_{\theta=0}; D_r(\tau) \triangleq \frac{\partial^{(r)}}{\partial\theta^{(r)}}\upsilon(\tau,\theta)\bigg|_{\theta=0}$$

are introduced so that the next theorem illustrates a tractable procedure of generating performance-measure statistics in time domain. This calculation is preferred to that of Eq. (3.17) for the reason that the resulting cumulant-generating equations now allow the incorporation of classes of linear feedback controllers in risk-averse tracking design synthesis.

Theorem 3.3.2 (Performance-Measure Statistics). *The tracking dynamics governed by Eqs. (3.4)–(3.5) attempt to follow the set-point signals $\gamma(t)$ and control input references $u_r(t)$ with the overtaking performance-measure (3.5). For $k \in \mathbb{Z}^+$, the kth performance-measure statistic is given by the closed-form*

$$\kappa_k = x_0^T H_k(t_0)x_0 + 2x_0^T \breve{D}_k(t_0) + D_k(t_0), \tag{3.18}$$

where the cumulant-generating variables $\{H_r(\tau)\}_{r=1}^k$, $\{\breve{D}_r(\tau)\}_{r=1}^k$ and $\{D_r(\tau)\}_{r=1}^k$ evaluated at $\tau = t_0$ satisfy the time-backward differential equations (with the dependence of $H_r(\tau)$, $\breve{D}_r(\tau)$ and $D_r(\tau)$ upon l_f and K suppressed)

$$\frac{d}{d\tau}H_1(\tau) = -[A(\tau)+B(\tau)K(\tau)]^T H_1(\tau) - H_1(\tau)[A(\tau)+B(\tau)K(\tau)] - Q(\tau) - K^T(\tau)R(\tau)K(\tau), \tag{3.19}$$

$$\frac{d}{d\tau}H_r(\tau) = -[A(\tau)+B(\tau)K(\tau)]^T H_r(\tau) - H_r(\tau)[A(\tau)+B(\tau)K(\tau)] - \sum_{s=1}^{r-1}\frac{2r!}{s!(r-s)!}H_s(\tau)G(\tau)WG^T(\tau)H_{r-s}(\tau), \tag{3.20}$$

$$\frac{d}{d\tau}\breve{D}_1(\tau) = -[A(\tau)+B(\tau)K(\tau)]^T \breve{D}_1(\tau) - H_1(\tau)B(\tau)\left[l_f(\tau)+u_r(\tau)\right] - K^T(\tau)R(\tau)l_f(\tau) + Q(\tau)\gamma(\tau), \tag{3.21}$$

$$\frac{d}{d\tau}\breve{D}_r(\tau) = -[A(\tau)+B(\tau)K(\tau)]^T \breve{D}_r(\tau) - H_r(\tau)B(\tau)\left[l_f(\tau)+u_r(\tau)\right], \tag{3.22}$$

$$\frac{d}{d\tau}D_1(\tau) = -\mathrm{Tr}\{H_1(\tau)G(\tau)WG^T(\tau)\} - 2\breve{D}_1^T(\tau)B(\tau)\left[l_f(\tau)+u_r(\tau)\right] - l_f^T(\tau)R(\tau)l_f(\tau) - \gamma^T(\tau)Q(\tau)\gamma(\tau), \quad (3.23)$$

$$\frac{d}{d\tau}D_r(\tau) = -\mathrm{Tr}\{H_r(\tau)G(\tau)WG^T(\tau)\} - 2\breve{D}_r^T(\tau)B(\tau)\left[l_f(\tau)+u_r(\tau)\right], \quad (3.24)$$

whereby the terminal-value conditions $H_1(t_f) = Q_f$, $H_r(t_f) = 0$ *for* $2 \le r \le k$; $\breve{D}_1(t_f) = -Q_f\gamma(t_f)$, $\breve{D}_r(t_f) = 0$ *for* $2 \le r \le k$, *and* $D_1(t_f) = \gamma^T(t_f)Q_f\gamma(t_f)$, $D_r(t_f) = 0$ *for* $2 \le r \le k$.

3.4 Statements of the Risk-Averse Control Problem

For the reason that the values of performance-measure statistics (3.18) depend in part of the known initial condition $x(t_0)$, the statistical optimal control is therefore dealt with a set of deterministic equations which describes the "trajectories" of the performance measure statistics (3.18). Henceforth, the linear-quadratic class of stochastic control problems with performance risk aversion is subsequently transformed into a deterministic control problem where the dynamics of the deterministic systems are described by a set of time-backward matrix and vector-valued differential equations (3.19)–(3.24).

In other words, the time-backward trajectories (3.19)–(3.24) are now considered as the "new" dynamical equations from which the resulting Mayer optimization [7] and associated value function in dynamic programming now depend on these "new" states $H_r(\tau)$, $\breve{D}_r(\tau)$, and $D_r(\tau)$, not the states $x(t)$ as traditionally expected. Furthermore, it is important to see that this mathematical representation (3.19)–(3.24) underlies the conceptual structure to extract the knowledge of intrinsic performance variability introduced by the process noise stochasticity in definite terms of performance-measure statistics (3.18).

Next it is convenient to introduce k-tuple variables $\mathscr{H}$, $\breve{\mathscr{D}}$, and $\mathscr{D}$ as follows $\mathscr{H}(\cdot) \triangleq (\mathscr{H}_1(\cdot),\ldots,\mathscr{H}_k(\cdot))$, $\breve{\mathscr{D}}(\cdot) \triangleq (\breve{\mathscr{D}}_1(\cdot),\ldots,\breve{\mathscr{D}}_k(\cdot))$ and $\mathscr{D}(\cdot) \triangleq (\mathscr{D}_1(\cdot),\ldots,\mathscr{D}_k(\cdot))$ for each element $\mathscr{H}_r \in \mathscr{C}^1([t_0,t_f];\mathbb{R}^{n\times n})$ of $\mathscr{H}$, $\breve{\mathscr{D}}_r \in \mathscr{C}^1([t_0,t_f];\mathbb{R}^n)$ of $\breve{\mathscr{D}}$, and $\mathscr{D}_r \in \mathscr{C}^1([t_0,t_f];\mathbb{R})$ of $\mathscr{D}$ having the representations $\mathscr{H}_r(\cdot) \triangleq H_r(\cdot)$, $\breve{\mathscr{D}}_r(\cdot) \triangleq \breve{D}_r(\cdot)$, and $\mathscr{D}_r(\cdot) \triangleq D_r(\cdot)$ with the right members satisfying the dynamic equations (3.19)–(3.24) on the finite horizon $[t_0,t_f]$.

The problem formulation is considerably simplified if the bounded and Lipschitz continuous mappings are introduced accordingly

$$\mathscr{F}_r : [t_0,t_f] \times (\mathbb{R}^{n\times n})^k \times \mathbb{R}^{m\times n} \mapsto \mathbb{R}^{n\times n},$$

$$\breve{\mathscr{G}}_r : [t_0,t_f] \times (\mathbb{R}^{n\times n})^k \times (\mathbb{R}^n)^k \times \mathbb{R}^{m\times n} \times \mathbb{R}^m \mapsto \mathbb{R}^n,$$

$$\mathscr{G}_r : [t_0,t_f] \times (\mathbb{R}^{n\times n})^k \times (\mathbb{R}^n)^k \times \mathbb{R}^m \mapsto \mathbb{R},$$

where the actions are given by

$$\mathscr{F}_1(\tau,\mathscr{H},K) \triangleq -[A(\tau)+B(\tau)K(\tau)]^T \mathscr{H}_1(\tau) - \mathscr{H}_1(\tau)[A(\tau)+B(\tau)K(\tau)] - Q(\tau) - K^T(\tau)R(\tau)K(\tau),$$

$$\mathscr{F}_r(\tau,\mathscr{H},K) \triangleq -[A(\tau)+B(\tau)K(\tau)]^T \mathscr{H}_r(\tau) - \mathscr{H}_r(\tau)[A(\tau)+B(\tau)K(\tau)] - \sum_{s=1}^{r-1} \frac{2r!}{s!(r-s)!} \mathscr{H}_s(\tau)G(\tau)WG^T(\tau)\mathscr{H}_{r-s}(\tau),$$

$$\breve{\mathscr{G}}_1(\tau,\mathscr{H},\breve{\mathscr{D}},K,l_f) \triangleq -[A(\tau)+B(\tau)K(\tau)]^T \breve{\mathscr{D}}_1(\tau) - \mathscr{H}_1(\tau)B(\tau)\left[l_f(\tau)+u_r(\tau)\right] - K^T(\tau)R(\tau)l_f(\tau) + Q(\tau)\gamma(\tau),$$

$$\breve{\mathscr{G}}_r(\tau,\mathscr{H},\breve{\mathscr{D}},K,l_f) \triangleq -[A(\tau)+B(\tau)K(\tau)]^T \breve{\mathscr{D}}_r(\tau) - \mathscr{H}_r(\tau)B(\tau)\left[l_f(\tau)+u_r(\tau)\right],$$

$$\mathscr{G}_1(\tau,\mathscr{H},\breve{\mathscr{D}},l_f) \triangleq -\mathrm{Tr}\{\mathscr{H}_1(\tau)G(\tau)WG^T(\tau)\} - 2\breve{\mathscr{D}}_1^T(\tau)B(\tau)\left[l_f(\tau)+u_r(\tau)\right] - l_f^T(\tau)R(\tau)l_f(\tau) - \gamma^T(\tau)Q(\tau)\gamma(\tau),$$

$$\mathscr{G}_r(\tau,\mathscr{H},\breve{\mathscr{D}},l_f) \triangleq -\mathrm{Tr}\{\mathscr{H}_r(\tau)G(\tau)WG^T(\tau)\} - 2\breve{\mathscr{D}}_r^T(\tau)B(\tau)\left[l_f(\tau)+u_r(\tau)\right].$$

For compactness of notations, the next product mappings are further needed

$$\mathscr{F}_1 \times \cdots \times \mathscr{F}_k : [t_0,t_f] \times (\mathbb{R}^{n\times n})^k \times \mathbb{R}^{m\times n} \mapsto (\mathbb{R}^{n\times n})^k,$$

$$\breve{\mathscr{G}}_1 \times \cdots \times \breve{\mathscr{G}}_k : [t_0,t_f] \times (\mathbb{R}^{n\times n})^k \times (\mathbb{R}^n)^k \times \mathbb{R}^{m\times n} \times \mathbb{R}^m \mapsto (\mathbb{R}^n)^k,$$

$$\mathscr{G}_1 \times \cdots \times \mathscr{G}_k : [t_0,t_f] \times (\mathbb{R}^{n\times n})^k \times (\mathbb{R}^n)^k \times \mathbb{R}^m \mapsto \mathbb{R}^k$$

along with the corresponding notations $\mathscr{F} \triangleq \mathscr{F}_1 \times \cdots \times \mathscr{F}_k$, $\breve{\mathscr{G}} \triangleq \breve{\mathscr{G}}_1 \times \cdots \times \breve{\mathscr{G}}_k$, and $\mathscr{G} \triangleq \mathscr{G}_1 \times \cdots \times \mathscr{G}_k$. Thus, the dynamic equations of motion (3.19)–(3.24) can be rewritten as follows

$$\frac{\mathrm{d}}{\mathrm{d}\tau}\mathscr{H}(\tau) = \mathscr{F}(\tau,\mathscr{H}(\tau),K(\tau)), \quad \mathscr{H}(t_f) \equiv \mathscr{H}_f,$$

$$\frac{\mathrm{d}}{\mathrm{d}\tau}\breve{\mathscr{D}}(\tau) = \breve{\mathscr{G}}(\tau,\mathscr{H}(\tau),\breve{\mathscr{D}}(\tau),K(\tau),l_f(\tau)), \quad \breve{\mathscr{D}}(t_f) \equiv \breve{\mathscr{D}}_f,$$

$$\frac{\mathrm{d}}{\mathrm{d}\tau}\mathscr{D}(\tau) = \mathscr{G}(\tau,\mathscr{H}(\tau),\breve{\mathscr{D}}(\tau),l_f(\tau)), \quad \mathscr{D}(t_f) \equiv \mathscr{D}_f,$$

whereby the k-tuple final values $\mathscr{H}_f \triangleq (Q_f,0,\ldots,0)$, $\breve{\mathscr{D}}_f \triangleq (-Q_f\gamma(t_f),0,\ldots,0)$, and $\mathscr{D}_f \triangleq (\gamma^T(t_f)Q_f\gamma(t_f),\ldots,0)$.

Notice that the product system uniquely determines $\mathscr{H}$, $\breve{\mathscr{D}}$, and $\mathscr{D}$ once the admissible affine input l_f and feedback gain K are specified. Hence, they are considered as $\mathscr{H} = \mathscr{H}(\cdot, K)$, $\breve{\mathscr{D}} = \breve{\mathscr{D}}(\cdot, K, l_f)$ and $\mathscr{D} = \mathscr{D}(\cdot, K, l_f)$. The risk-averse performance index is defined by these control parameters l_f and K.

Definition 3.4.1 (Risk-Value Aware Performance Index). Fix $k \in \mathbb{Z}^+$ and the sequence $\mu = \{\mu_r \geq 0\}_{r=1}^k$ with $\mu_1 > 0$. Then, for the given (t_0, x_0), the risk-value aware performance index of Mayer type, i.e., $\phi_0^{tk} : \{t_0\} \times (\mathbb{R}^{n\times n})^k \times (\mathbb{R}^n)^k \times \mathbb{R}^k \mapsto \mathbb{R}^+$ for the generalized tracking problem is defined by

$$\phi_0^{tk}\left(t_0, \mathscr{H}(t_0), \breve{\mathscr{D}}(t_0), \mathscr{D}(t_0)\right) \triangleq \underbrace{\mu_1 \kappa_1}_{\text{Value Measure}} + \underbrace{\mu_2 \kappa_2 + \cdots + \mu_k \kappa_k}_{\text{Risk Measures}}$$

$$= \sum_{r=1}^{k} \mu_r [x_0^T \mathscr{H}_r(t_0) x_0 + 2x_0^T \breve{\mathscr{D}}_r(t_0) + \mathscr{D}_r(t_0)]. \quad (3.25)$$

The real constant scalars μ_r represent different degrees of freedom to shape the distribution of closed-loop tracking performance wherever they matter the most by a means of placing particular weights on any specific performance-measure statistics (i.e., mean, variance, skewness, flatness, etc.) associated with the performance measure (3.5). The unique solutions $\{\mathscr{H}_r(\cdot)\}_{r=1}^k$, $\left\{\breve{\mathscr{D}}_r(\cdot)\right\}_{r=1}^k$ and $\{\mathscr{D}_r(\cdot)\}_{r=1}^k$ evaluated at $\tau = t_0$ satisfy the time-backward equations of motion

$$\frac{d}{d\tau}\mathscr{H}(\tau) = \mathscr{F}(\tau, \mathscr{H}(\tau), K(\tau)), \quad \mathscr{H}(t_f), \quad (3.26)$$

$$\frac{d}{d\tau}\breve{\mathscr{D}}(\tau) = \breve{\mathscr{G}}\left(\tau, \mathscr{H}(\tau), \breve{\mathscr{D}}(\tau), K(\tau), l_f(\tau)\right), \quad \breve{\mathscr{D}}(t_f), \quad (3.27)$$

$$\frac{d}{d\tau}\mathscr{D}(\tau) = \mathscr{G}\left(\tau, \mathscr{H}(\tau), \breve{\mathscr{D}}(\tau), l_f(\tau)\right), \quad \mathscr{D}(t_f). \quad (3.28)$$

For the given terminal data $(t_f, \mathscr{H}_f, \breve{\mathscr{D}}_f, \mathscr{D}_f)$, the class of feedback controls whose admissible affine inputs and feedback gains are next defined is only admitted herein.

Definition 3.4.2 (Admissible Inputs and Feedback Gains). Let compact subsets $\overline{L} \subset \mathbb{R}^m$ and $\overline{K} \subset \mathbb{R}^{m\times n}$ be the sets of allowable linear control inputs and gain values. For the given $k \in \mathbb{Z}^+$ and the sequence $\mu = \{\mu_r \geq 0\}_{r=1}^k$ with $\mu_1 > 0$, the set of admissible affine inputs $\mathscr{L}_{t_f, \mathscr{H}_f, \breve{\mathscr{D}}_f, \mathscr{D}_f; \mu}$ and feedback gains $\mathscr{K}_{t_f, \mathscr{H}_f, \breve{\mathscr{D}}_f, \mathscr{D}_f; \mu}$ are respectively assumed to be the classes of $\mathscr{C}([t_0, t_f]; \mathbb{R}^m)$ and $\mathscr{C}([t_0, t_f]; \mathbb{R}^{m\times n})$ with values $l_f(\cdot) \in \overline{L}$ and $K(\cdot) \in \overline{K}$ for which solutions to the dynamic equations (3.26)–(3.28) exist on the interval of optimization $[t_0, t_f]$.

As before, the mathematical mechanisms for the dynamic programming approach which is adapted specifically for the Mayer optimization problem within the context of statistical optimal control are stated and the outline of the proofs will be given.

Definition 3.4.3 (Optimization Problem of Mayer Type). Suppose that $k \in \mathbb{Z}^+$ and the sequence $\mu = \{\mu_r \geq 0\}_{r=1}^k$ with $\mu_1 > 0$ are fixed. Then, the Mayer optimization problem over $[t_0, t_f]$ is given by the minimization of the risk-value aware performance index (3.25) over $l_f(\cdot) \in \mathscr{L}_{t_f,\mathscr{H}_f,\breve{\mathscr{D}}_f,\mathscr{D}_f;\mu}$, $K(\cdot) \in \mathscr{K}_{t_f,\mathscr{H}_f,\breve{\mathscr{D}}_f,\mathscr{D}_f;\mu}$ and subject to the dynamic equations of motion (3.26)–(3.28).

The subsequent results will then illustrate a construction of potential candidates for the value function.

Definition 3.4.4 (Reachable Set). Let reachable set $\mathscr{Q} \triangleq \Big\{(\varepsilon, \mathscr{Y}, \breve{\mathscr{Z}}, \mathscr{Z}) \in [t_0, t_f] \times (\mathbb{R}^{n\times n})^k \times (\mathbb{R}^n)^k \times \mathbb{R}^k$ such that $\mathscr{L}_{\varepsilon,\mathscr{Y},\breve{\mathscr{Z}},\mathscr{Z};\mu} \times \mathscr{K}_{\varepsilon,\mathscr{Y},\breve{\mathscr{Z}},\mathscr{Z};\mu} \neq \emptyset\Big\}$.

By adapting to the initial cost problem and the terminologies present in the risk-averse control, the Hamilton-Jacobi-Bellman (HJB) equation satisfied by the value function $\mathscr{V}(\varepsilon, \mathscr{Y}, \breve{\mathscr{Z}}, \mathscr{Z})$ is given as follows.

Theorem 3.4.1 (HJB Equation for Mayer Problem). *Let $(\varepsilon, \mathscr{Y}, \breve{\mathscr{Z}}, \mathscr{Z})$ be any interior point of the reachable set $\mathscr{Q}$ at which the value function $\mathscr{V}(\varepsilon, \mathscr{Y}, \breve{\mathscr{Z}}, \mathscr{Z})$ is differentiable. If there exist optimal affine signal $l_f^* \in \mathscr{L}_{\varepsilon,\mathscr{Y},\breve{\mathscr{Z}},\mathscr{Z};\mu}$ and feedback gain $K^* \in \mathscr{K}_{\varepsilon,\mathscr{Y},\breve{\mathscr{Z}},\mathscr{Z};\mu}$, then the partial differential equation of dynamic programming*

$$
\begin{aligned}
0 = \min_{l_f \in \overline{L}, K \in \overline{K}} \Big\{ & \frac{\partial}{\partial \varepsilon} \mathscr{V}(\varepsilon, \mathscr{Y}, \breve{\mathscr{Z}}, \mathscr{Z}) + \frac{\partial}{\partial \operatorname{vec}(\mathscr{Y})} \mathscr{V}(\varepsilon, \mathscr{Y}, \breve{\mathscr{Z}}, \mathscr{Z}) \operatorname{vec}(\mathscr{F}(\varepsilon, \mathscr{Y}, K)) \\
& + \frac{\partial}{\partial \operatorname{vec}(\breve{\mathscr{Z}})} \mathscr{V}(\varepsilon, \mathscr{Y}, \breve{\mathscr{Z}}, \mathscr{Z}) \operatorname{vec}(\breve{\mathscr{G}}(\varepsilon, \mathscr{Y}, \breve{\mathscr{Z}}, K, l_f)) \\
& + \frac{\partial}{\partial \operatorname{vec}(\mathscr{Z})} \mathscr{V}(\varepsilon, \mathscr{Y}, \breve{\mathscr{Z}}, \mathscr{Z}) \operatorname{vec}(\mathscr{G}(\varepsilon, \mathscr{Y}, \breve{\mathscr{Z}}, l_f)) \Big\}
\end{aligned}
\tag{3.29}
$$

is satisfied with the boundary condition governed by

$$\mathscr{V}(t_0, \mathscr{H}(t_0), \breve{\mathscr{D}}(t_0), \mathscr{D}(t_0)) = \phi_0^{tk}(t_0, \mathscr{H}(t_0), \breve{\mathscr{D}}(t_0), \mathscr{D}(t_0)).$$

Proof. By what has shown by [8], the proof for the result here is readily proven. □

Next another approach to sufficiency is the verification result which is then utilized to verify admissible feedback controls with performance risk aversion.

Theorem 3.4.2 (Verification Theorem). *Fix $k \in \mathbb{Z}^+$ and let $\mathscr{W}(\varepsilon, \mathscr{Y}, \breve{\mathscr{Z}}, \mathscr{Z})$ be a continuously differentiable solution of the HJB equation (3.29) which satisfies the boundary condition*

$$\mathscr{W}(t_0, \mathscr{H}_0, \breve{\mathscr{D}}_0, \mathscr{D}_0) = \phi_0^{tk}(t_0, \mathscr{H}_0, \breve{\mathscr{D}}_0, \mathscr{D}_0). \tag{3.30}$$

Let $(t_f, \mathscr{H}_f, \breve{\mathscr{D}}_f, \mathscr{D}_f)$ be in $\mathscr{Q}$; (l_f, K) in $\mathscr{L}_{t_f,\mathscr{H}_f,\breve{\mathscr{D}}_f,\mathscr{D}_f;\mu} \times \mathscr{K}_{t_f,\mathscr{H}_f,\breve{\mathscr{D}}_f,\mathscr{D}_f;\mu}$; $\mathscr{H}$, $\breve{\mathscr{D}}$ and $\mathscr{D}$ the corresponding solutions of Eqs. (3.26)–(3.28). Then, $\mathscr{W}(\tau, \mathscr{H}(\tau), \breve{\mathscr{D}}(\tau), \mathscr{D}(\tau))$ is a time-backward increasing function of τ. If (l_f^*, K^*) is *in* $\mathscr{L}_{t_f,\mathscr{H}_f,\check{\mathscr{D}}_f,\mathscr{D}_f;\mu} \times \mathscr{K}_{t_f,\mathscr{H}_f,\check{\mathscr{D}}_f,\mathscr{D}_f;\mu}$ *defined on* $[t_0,t_f]$ *with corresponding solutions,* $\mathscr{H}^*$, $\check{\mathscr{D}}^*$, *and* $\mathscr{D}^*$ *of Eqs. (3.26)–(3.28) such that for* $\tau \in [t_0,t_f]$

$$
\begin{aligned}
0 &= \frac{\partial}{\partial \varepsilon}\mathscr{W}\left(\tau,\mathscr{H}^*(\tau),\check{\mathscr{D}}^*(\tau),\mathscr{D}^*(\tau)\right) \\
&+ \frac{\partial}{\partial \mathrm{vec}(\mathscr{Y})}\mathscr{W}\left(\tau,\mathscr{H}^*(\tau),\check{\mathscr{D}}^*(\tau),\mathscr{D}^*(\tau)\right)\mathrm{vec}\left(\mathscr{F}\left(\tau,\mathscr{H}^*(\tau),K^*(\tau)\right)\right) \\
&+ \frac{\partial}{\partial \mathrm{vec}(\check{\mathscr{Z}})}\mathscr{W}\left(\tau,\mathscr{H}^*(\tau),\check{\mathscr{D}}^*(\tau),\mathscr{D}^*(\tau)\right)\mathrm{vec}\left(\check{\mathscr{G}}\left(\tau,\mathscr{H}^*(\tau),\check{\mathscr{D}}^*(\tau),K^*(\tau),l_f^*(\tau)\right)\right) \\
&+ \frac{\partial}{\partial \mathrm{vec}(\mathscr{Z})}\mathscr{W}\left(\tau,\mathscr{H}^*(\tau),\check{\mathscr{D}}^*(\tau),\mathscr{D}^*(\tau)\right)\mathrm{vec}\left(\mathscr{G}\left(\tau,\mathscr{H}^*(\tau),\check{\mathscr{D}}^*(\tau),l_f^*(\tau)\right)\right)
\end{aligned}
\tag{3.31}
$$

then both l_f^* *and* K^* *are optimal. Moreover, it follows that*

$$
\mathscr{W}\left(\varepsilon,\mathscr{Y},\check{\mathscr{Z}},\mathscr{Z}\right) = \mathscr{V}\left(\varepsilon,\mathscr{Y},\check{\mathscr{Z}},\mathscr{Z}\right),
\tag{3.32}
$$

where $\mathscr{V}\left(\varepsilon,\mathscr{Y},\check{\mathscr{Z}},\mathscr{Z}\right)$ *is the value function.*

Proof. With the aid of the recent development [7, 8], the proof then follows for the results stated here. □

3.5 Optimal Risk-Averse Tracking Solution

It is obvious that the optimization problem considered herein is in "Mayer form". It is therefore solved by applying an adaptation of the Mayer-form verification theorem of dynamic programming given in [7]. In the language of dynamic programming, it requires to parameterize all starting times and states of a family of optimization problems as $\left(\varepsilon,\mathscr{Y},\check{\mathscr{Z}},\mathscr{Z}\right)$. For instance, the states of the system (3.26)–(3.28) defined on $[t_0,\varepsilon]$ with the terminal values are now denoted by $\mathscr{H}(\varepsilon)\equiv\mathscr{Y}$, $\check{\mathscr{D}}(\varepsilon)\equiv\check{\mathscr{Z}}$, and $\mathscr{D}(\varepsilon)\equiv\mathscr{Z}$. Next the choice of a candidate solution to the HJB equation (3.29) depends on properties like the differentiability and quadratic-affine nature of the risk-value aware performance index (3.25), e.g.,

$$
\begin{aligned}
\mathscr{W}\left(\varepsilon,\mathscr{Y},\check{\mathscr{Z}},\mathscr{Z}\right) &= x_0^T \sum_{r=1}^{k} \mu_r\left(\mathscr{Y}_r + \mathscr{E}_r(\varepsilon)\right)x_0 \\
&\quad + 2x_0^T \sum_{r=1}^{k} \mu_r\left(\check{\mathscr{Z}}_r + \check{\mathscr{T}}_r(\varepsilon)\right) + \sum_{r=1}^{k} \mu_r\left(\mathscr{Z}_r + \mathscr{T}_r(\varepsilon)\right),
\end{aligned}
\tag{3.33}
$$

whereby the time-parametric functions $\mathscr{E}_r \in \mathscr{C}^1([t_0,t_f];\mathbb{R}^{n\times n})$, $\check{\mathscr{T}}_r \in \mathscr{C}^1([t_0,t_f];\mathbb{R}^n)$ and $\mathscr{T}_r \in \mathscr{C}^1([t_0,t_f];\mathbb{R})$ are yet to be determined.

Therefore, one can obtain the derivative of $\mathscr{W}\left(\varepsilon,\mathscr{Y},\breve{\mathscr{Z}},\mathscr{Z}\right)$ with respect to ε

$$\begin{aligned}\frac{d}{d\varepsilon}\mathscr{W}\left(\varepsilon,\mathscr{Y},\breve{\mathscr{Z}},\mathscr{Z}\right) =& x_0^T\sum_{r=1}^{k}\mu_r\left(\mathscr{F}_r(\varepsilon,\mathscr{Y},K)+\frac{d}{d\varepsilon}\mathscr{E}_r(\varepsilon)\right)x_0\\ &+2x_0^T\sum_{r=1}^{k}\mu_r\left(\breve{\mathscr{G}}_r\left(\varepsilon,\mathscr{Y},\breve{\mathscr{Z}},K,l_f\right)+\frac{d}{d\varepsilon}\breve{\mathscr{T}}_r(\varepsilon)\right)\\ &+\sum_{r=1}^{k}\mu_r\left(\mathscr{G}_r(\varepsilon,\mathscr{Y},\breve{\mathscr{Z}},l_f)+\frac{d}{d\varepsilon}\mathscr{T}_r(\varepsilon)\right)\end{aligned} \tag{3.34}$$

provided that $l_f\in\overline{L}$ and $K\in\overline{K}$. Trying this candidate for the value function (3.33) into the HJB equation (3.29) yields

$$\begin{aligned}0\equiv\min_{l_f\in\overline{L},K\in\overline{K}}&\left\{x_0^T\sum_{r=1}^{k}\mu_r\left(\mathscr{F}_r(\varepsilon,\mathscr{Y},K)+\frac{d}{d\varepsilon}\mathscr{E}_r(\varepsilon)\right)x_0\right.\\ &+2x_0^T\sum_{r=1}^{k}\mu_r\left(\breve{\mathscr{G}}_r(\varepsilon,\mathscr{Y},\breve{\mathscr{Z}},K,l_f)+\frac{d}{d\varepsilon}\breve{\mathscr{T}}_r(\varepsilon)\right)\\ &\left.+\sum_{r=1}^{k}\mu_r(\mathscr{G}_r\left(\varepsilon,\mathscr{Y},\breve{\mathscr{Z}},l_f)+\frac{d}{d\varepsilon}\mathscr{T}_r(\varepsilon)\right)\right\}.\end{aligned} \tag{3.35}$$

Since the initial condition x_0 is an arbitrary vector, the necessary condition for an extremum of Eq. (3.25) on $[t_0,\varepsilon]$ is obtained by differentiating the expression within the bracket of Eq. (3.35) with respect to the control parameters l_f and K as follows

$$l_f(\varepsilon,\breve{\mathscr{Z}})=-R^{-1}(\varepsilon)B^T(\varepsilon)\sum_{r=1}^{k}\hat{\mu}_r\breve{\mathscr{Z}}_r, \tag{3.36}$$

$$K(\varepsilon,\mathscr{Y})=-R^{-1}(\varepsilon)B^T(\varepsilon)\sum_{r=1}^{k}\hat{\mu}_r\mathscr{Y}_r, \tag{3.37}$$

where the weightings $\hat{\mu}_r\triangleq\mu_i/\mu_1$ are normalized by $\mu_1>0$.

What remains is to exhibit the time parametric functions for the candidate function $\mathscr{W}\left(\varepsilon,\mathscr{Y},\breve{\mathscr{Z}},\mathscr{Z}\right)$ of the value function, i.e., $\{\mathscr{E}_r(\cdot)\}_{r=1}^k$, $\left\{\breve{\mathscr{T}}_r(\cdot)\right\}_{r=1}^k$, and $\{\mathscr{T}_r(\cdot)\}_{r=1}^k$ which yield a sufficient condition to have the left-hand side of Eq. (3.35) being zero for any $\varepsilon\in[t_0,t_f]$, when $\{\mathscr{Y}_r\}_{r=1}^k$, $\left\{\breve{\mathscr{Z}}_r\right\}_{r=1}^k$ and $\{\mathscr{Z}_r\}_{r=1}^k$ are evaluated along the solutions of the cumulant-generating equations (3.26)–(3.28).

With a careful examination of the right-hand side of Eq. (3.35), $\{\mathscr{E}_r(\cdot)\}_{r=1}^k$, $\left\{\breve{\mathscr{T}}_r(\cdot)\right\}_{r=1}^k$ and $\{\mathscr{T}_r(\cdot)\}_{r=1}^k$ selectively satisfy the following time-forward differential equations

$$\begin{aligned}
\frac{d}{d\varepsilon}\mathscr{E}_1(\varepsilon) = & A^T(\varepsilon)\mathscr{H}_1(\varepsilon) + \mathscr{H}_1(\varepsilon)A(\varepsilon) + Q(\varepsilon) \\
& - \mathscr{H}_1(\varepsilon)B(\varepsilon)R^{-1}(\varepsilon)B^T(\varepsilon)\sum_{s=1}^{k}\hat{\mu}_s\mathscr{H}_s(\varepsilon) \\
& - \sum_{s=1}^{k}\hat{\mu}_s\mathscr{H}_s(\varepsilon)B(\varepsilon)R^{-1}(\varepsilon)B^T(\varepsilon)\mathscr{H}_1(\varepsilon) \\
& + \sum_{r=1}^{k}\hat{\mu}_r\mathscr{H}_r(\varepsilon)B(\varepsilon)R^{-1}(\varepsilon)B^T(\varepsilon)\sum_{s=1}^{k}\hat{\mu}_s\mathscr{H}_s(\varepsilon),
\end{aligned} \tag{3.38}$$

$$\begin{aligned}
\frac{d}{d\varepsilon}\mathscr{E}_r(\varepsilon) = & A^T(\varepsilon)\mathscr{H}_r(\varepsilon) + \mathscr{H}_r(\varepsilon)A(\varepsilon) - \mathscr{H}_r(\varepsilon)B(\varepsilon)R^{-1}(\varepsilon)B^T(\varepsilon)\sum_{s=1}^{k}\hat{\mu}_s\mathscr{H}_s(\varepsilon) \\
& - \sum_{s=1}^{k}\hat{\mu}_s\mathscr{H}_s(\varepsilon)B(\varepsilon)R^{-1}(\varepsilon)B^T(\varepsilon)\mathscr{H}_r(\varepsilon) \\
& + \sum_{s=1}^{r-1}\frac{2r!}{s!(r-s)!}\mathscr{H}_s(\varepsilon)G(\varepsilon)WG^T(\varepsilon)\mathscr{H}_{r-s}(\varepsilon),
\end{aligned} \tag{3.39}$$

$$\begin{aligned}
\frac{d}{d\varepsilon}\breve{\mathscr{T}}_1(\varepsilon) = & A^T(\varepsilon)\breve{\mathscr{D}}_1(\varepsilon) - Q(\varepsilon)\gamma(\varepsilon) - \sum_{s=1}^{k}\hat{\mu}_s\mathscr{H}_s(\varepsilon)B(\varepsilon)R^{-1}(\varepsilon)B^T(\varepsilon)\breve{\mathscr{D}}_1(\varepsilon) \\
& + \mathscr{H}_1(\varepsilon)B(\varepsilon)[-R^{-1}(\varepsilon)B^T(\varepsilon)\sum_{s=1}^{k}\hat{\mu}_s\breve{\mathscr{D}}_r(\varepsilon) + u_r(\varepsilon)] \\
& + \sum_{s=1}^{k}\hat{\mu}_s\mathscr{H}_s(\varepsilon)B(\varepsilon)R^{-1}(\varepsilon)B^T(\varepsilon)\sum_{s=1}^{k}\hat{\mu}_s\mathscr{D}_s(\varepsilon),
\end{aligned} \tag{3.40}$$

$$\begin{aligned}
\frac{d}{d\varepsilon}\breve{\mathscr{T}}_r(\varepsilon) = & A^T(\varepsilon)\breve{\mathscr{D}}_r(\varepsilon) - \sum_{s=1}^{k}\hat{\mu}_s\mathscr{H}_s(\varepsilon)B(\varepsilon)R^{-1}(\varepsilon)B^T(\varepsilon)\breve{\mathscr{D}}_r(\varepsilon) \\
& + \mathscr{H}_r(\varepsilon)B(\varepsilon)\left[-R^{-1}(\varepsilon)B^T(\varepsilon)\sum_{s=1}^{k}\hat{\mu}_s\breve{\mathscr{D}}_s(\varepsilon) + u_r(\varepsilon)\right],
\end{aligned} \tag{3.41}$$

$$\begin{aligned}
\frac{d}{d\varepsilon}\mathscr{T}_1(\varepsilon) = & \mathrm{Tr}\left\{\mathscr{H}_1(\varepsilon)G(\varepsilon)WG^T(\varepsilon)\right\} + \gamma^T(\varepsilon)Q(\varepsilon)\gamma(\varepsilon) \\
& + 2\breve{\mathscr{D}}_1^T(\varepsilon)B(\varepsilon)\left[-R^{-1}(\varepsilon)B^T(\varepsilon)\sum_{s=1}^{k}\hat{\mu}_s\breve{\mathscr{D}}_s(\varepsilon) + u_r(\varepsilon)\right] \\
& + \sum_{s=1}^{k}\hat{\mu}_s\breve{\mathscr{D}}_s^T(\varepsilon)B(\varepsilon)R^{-1}(\varepsilon)B^T(\varepsilon)\sum_{s=1}^{k}\hat{\mu}_s\breve{\mathscr{D}}_s(\varepsilon),
\end{aligned} \tag{3.42}$$

$$\frac{d}{d\varepsilon}\mathscr{T}_r(\varepsilon) = \mathrm{Tr}\left\{\mathscr{H}_r(\varepsilon)G(\varepsilon)WG^T(\varepsilon)\right\}$$
$$+2\breve{\mathscr{D}}_r^T(\varepsilon)B(\varepsilon)\left[-R^{-1}(\varepsilon)B^T(\varepsilon)\sum_{s=1}^{k}\hat{\mu}_s\breve{\mathscr{D}}_s(\varepsilon)+u_r(\varepsilon)\right]. \qquad (3.43)$$

The boundary condition of $\mathscr{W}(\varepsilon,\mathscr{Y},\breve{\mathscr{Z}},\mathscr{Z})$ implies that

$$x_0^T\sum_{r=1}^{k}\mu_r\left(\mathscr{H}_r(t_0)+\mathscr{E}_r(t_0)\right)x_0+2x_0^T\sum_{r=1}^{k}\mu_r\left(\breve{\mathscr{D}}_r(t_0)+\breve{\mathscr{T}}_r(t_0)\right)$$
$$+\sum_{r=1}^{k}\mu_r\left(\mathscr{D}_r(t_0)+\mathscr{T}_r(t_0)\right)=x_0^T\sum_{r=1}^{k}\mu_r\mathscr{H}_r(t_0)x_0+2x_0^T\sum_{r=1}^{k}\mu_r\breve{\mathscr{D}}_r(t_0)+\sum_{r=1}^{k}\mu_r\mathscr{D}_r(t_0).$$

The initial-value conditions for the time-parametric equations (3.38)–(3.43) are given as follows $\mathscr{E}_r(t_0)=0$, $\breve{\mathscr{T}}_r(t_0)=0$ and $\mathscr{T}_r(t_0)=0$. Finally, the optimal linear input (3.36) and feedback gain (3.37) minimizing the performance index (3.25) become optimal

$$l_f^*(\varepsilon)=-R^{-1}(\varepsilon)B^T(\varepsilon)\sum_{r=1}^{k}\hat{\mu}_r\breve{\mathscr{D}}_r^*(\varepsilon),$$
$$K^*(\varepsilon)=-R^{-1}(\varepsilon)B^T(\varepsilon)\sum_{r=1}^{k}\hat{\mu}_r\mathscr{H}_r^*(\varepsilon).$$

Theorem 3.5.1 (Optimal Risk-Averse Tracking Solution). *Suppose (A,B) is uniformly stabilizable and (C,A) is uniformly detectable where $C^T(t)C(t)\triangleq Q(t)$. Assume further $k\in\mathbb{Z}^+$ and the sequence $\mu=\{\mu_r\ge 0\}_{r=1}^k$ with $\mu_1>0$ fixed. The risk-averse tracking solution for the generalized tracking problem whose the state dynamics $x(t)$ and the control input $u(t)$ governed by Eqs. (3.1) and (3.2) will track closely the desired trajectory $\gamma(t)$ and reference command $u_r(t)$, is given by the risk-averse control policy*

$$u^*(t)=K^*(t)x^*(t)+l_f^*(t)+p(t),\quad t=t_0+t_f-\tau, \qquad (3.44)$$
$$K^*(\tau)=-R^{-1}(\tau)B^T(\tau)\sum_{r=1}^{k}\hat{\mu}_r\mathscr{H}_r^*(\tau),\quad \tau\in[t_0,t_f], \qquad (3.45)$$
$$l_f^*(\tau)=-R^{-1}(\tau)B^T(\tau)\sum_{r=1}^{k}\hat{\mu}_r\breve{\mathscr{D}}_r^*(\tau), \qquad (3.46)$$

where the normalized weightings $\hat{\mu}_r\triangleq\mu_i/\mu_1$ emphasize on different design freedom of shaping the probability density function of the generalized performance-measure (3.5) of the chi-squared type. The optimal solutions $\{\mathscr{H}_r^*(\tau)\}_{r=1}^k$ are *satisfying the backward-in-time matrix-valued differential equations*

$$\frac{d}{d\tau}\mathscr{H}_1^*(\tau) = -[A(\tau)+B(\tau)K^*(\tau)]^T \mathscr{H}_1^*(\tau) - \mathscr{H}_1^*(\tau)[A(\tau)+B(\tau)K^*(\tau)] - Q(\tau) - K^{*T}(\tau)R(\tau)K^*(\tau), \qquad \mathscr{H}_1^*(t_f) = Q_f \tag{3.47}$$

and, for $2 \le r \le k$

$$\frac{d}{d\tau}\mathscr{H}_r^*(\tau) = -[A(\tau)+B(\tau)K^*(\tau)]^T \mathscr{H}_r^*(\tau) - \mathscr{H}_r^*(\tau)[A(\tau)+B(\tau)K^*(\tau)] - \sum_{s=1}^{r-1}\frac{2r!}{s!(r-s)!}\mathscr{H}_s^*(\tau)G(\tau)WG^T(\tau)\mathscr{H}_{r-s}^*(\tau), \; \mathscr{H}_r^*(t_f) = 0. \tag{3.48}$$

In addition, the optimal solutions $\{\breve{\mathscr{D}}_r^*(\tau)\}_{r=1}^k$ *are solving the backward-in-time vector-valued differential equations*

$$\frac{d}{d\tau}\breve{\mathscr{D}}_1^*(\tau) = -[A(\tau)+B(\tau)K^*(\tau)]^T \breve{\mathscr{D}}_1^*(\tau) - \mathscr{H}_1^*(\tau)B(\tau)\left[l_f^*(\tau)+u_r(\tau)\right] - K^{*T}(\tau)R(\tau)l_f^*(\tau) + Q(\tau)\gamma(\tau), \qquad \breve{\mathscr{D}}_1^*(t_f) = -Q_f\gamma(t_f) \tag{3.49}$$

and, for $2 \le r \le k$

$$\frac{d}{d\tau}\breve{\mathscr{D}}_r^*(\tau) = -[A(\tau)+B(\tau)K^*(\tau)]^T \breve{\mathscr{D}}_r^*(\tau) - \mathscr{H}_r^*(\tau)B(\tau)\left[l_f^*(\tau)+u_r(\tau)\right], \quad \breve{\mathscr{D}}_r^*(t_f) = 0. \tag{3.50}$$

Notice that the optimal feedback gain (3.45) and affine control input (3.46) operate dynamically on the time-backward histories of the cumulant-supporting equations (3.47)–(3.48) and (3.49)–(3.50) from the final to the current time. Moreover, it is important to see that these dynamical equations are functions of the noise process characteristics, i.e., second-order statistic W. Hence, the high confident tracking paradigm consisting optimal feedback gain (3.45) and affine input (3.46) has traded the certainty equivalence property, as one may normally obtain from the special case of traditional linear-quadratic tracking, for the adaptability to deal with uncertain environments and performance variations.

3.6 Chapter Summary

The present chapter proposes an advanced solution concept and a novel paradigm of designing feedback controls for a class of stochastic systems to simultaneously track reference trajectory and command input in accordance of the so-called, risk-averse performance index that is now composed of multiple selective performance-measure statistics beyond the traditional statistical average. A numerical procedure of calculating higher-order statistics associated with the chi-squared performance-measure is also obtained. The robustness and uncertainty of tracking performance is therefore maintained compactly and robustly. The complexity of the feedback controller may however increase considerably, depending on how many performance-measure statistics of the target probability density function are to be optimized.

References

1. Artstein, Z., Leizarowitz, A.: Tracking periodic signals with the overtaking criterion. IEEE Trans. Automat. Contr. **30**, 1123–1126 (1985)
2. Leizarowitz, A.: Tracking nonperiodic trajectories with the overtaking criterion. Appl. Math. Optim. **14**, 155–171 (1986)
3. Tan, H., Rugh, W.J.: On overtaking optimal tracking for linear systems. Syst. Contr. Lett. **33**, 63–72 (1998)
4. Pham, K.D.: Cost cumulant-based control for a class of linear-quadratic tracking problems. In: Proceedings of the American Control Conference, pp. 335-340 (2007)
5. Field, R.V., Bergman, L.A.: Reliability based approach to linear covariance control design. J. Eng. Mech. **124**(2), 193–199 (1998)
6. Hansen, L.P., Sargent, T.J., Tallarini, T.D. Jr.: Robust permanent income and pricing. Rev. Econ. Stud. **66**, 873–907 (1999)
7. Fleming, W.H., Rishel, R.W.: Deterministic and Stochastic Optimal Control. Springer, New York (1975)
8. Pham, K.D.: Performance-reliability aided decision making in multiperson quadratic decision games against jamming and estimation confrontations. In: Giannessi, F. (ed.) J. Optim. Theor. Appl. **149**(3), 559–629 (2011)

Chapter 4
Performance Risk Management in Servo Systems

Abstract This chapter provides a concise and up-to-date analysis of the foundations of performance robustness of a linear-quadratic class of servo systems with respect to variability in a stochastic environment. The dynamics of servo systems are corrupted by a standard stationary Wiener process and include input functions that are controlled by statistical optimal controllers. Basic assumptions are that the controllers have access to the current value of the states of the systems and are capable of learning about performance uncertainty of the systems that are now affected by stochastic elements, e.g., model deviations and exogenous disturbances. The controller considered here optimizes a multi-objective criterion over time where optimization takes place with high regard for sample realizations by the stochastic elements mentioned above. It is found that the optimal servo in the finite-horizon case is a novel two-degrees-of-freedom controller with: one, a feedback controller with state measurements that is robust against performance uncertainty; two, a model-following controller that minimizes the difference between the reference model and the system outputs.

4.1 Introduction

In recent works [1–4], the statistical optimal control for a class of optimal stochastic regulator problems has been developed for the task of returning systems to either zero or prespecified states in a complete statistical description of chi-squared random measures of performance. In fact, this regulator problem class is a special case of a wider class of problems where it is required that the outputs of a system follow a reference signal which in turn belongs to a known class of signals. The present research investigation now examines a possible extension of this generalized stochastic regulator theory developed so far in such a way that the resulting servo controller consists of a feedback controller together with a model-following controller involving processing of the desired reference signals to ensure the outputs of a linear stochastic system follow as closely as possible the outputs of

K.D. Pham, *Linear-Quadratic Controls in Risk-Averse Decision Making*,
SpringerBriefs in Optimization, DOI 10.1007/978-1-4614-5079-5_4,

a reference system in accordance of a given target probability density function of chi-squared random measure of performance. To the best knowledge of the author, the theoretical development in the sequel appears to be the first of its kind and its novel solution concepts are related well with the extensive literature on tracking and feedforward control design problems such as the earlier works of [5–7], just to name a few. Most of these works only considered the traditional measure of average performance using dynamic programming approach.

The results here will effectively address two key issues that have not been dealt with so far in the stochastic control literature, namely, how to quantify higher-order characteristics of performance uncertainty with respect to all sample realizations of the underlying stochastic process and how to design new model-following strategies that directly influence the performance distribution and robustness. Therefore, the enabling solution included here will bring one step closer to the realization of optimal tracking of stochastic systems with multi-attribute performance guarantees.

4.2 The Performance Information Process

As noted in Chaps. 2 and 3, the design of a control system with performance risk consequences involves the design of a performance information process which provides data for the decisions taken by risk-averse control designers. Henceforth, most of the development in this section focuses on the performance information process. For now, a class of servo stochastic systems whose state and output dynamics modeled on $[t_0, t_f]$ are given by

$$dx(t) = (A(t)x(t) + B(t)u(t))\,dt + G(t)dw(t)\,, \quad x(t_0), \tag{4.1}$$

$$z(t) = C(t)x(t), \tag{4.2}$$

whereby the continuous-time coefficients $A \in \mathscr{C}([t_0,t_f];\mathbb{R}^{n\times n})$, $B \in \mathscr{C}([t_0,t_f];\mathbb{R}^{n\times m})$, $C \in \mathscr{C}([t_0,t_f];\mathbb{R}^{r\times n})$, and $G \in \mathscr{C}([t_0,t_f];\mathbb{R}^{n\times p})$. The stochastic elements $w(t) \triangleq w(t,\omega) : [t_0,t_f] \times \Omega \mapsto \mathbb{R}^p$ is the p-dimensional Wiener process starting from t_0 with action space of Ω, which is independent of the known $x(t_0) \triangleq x_0$. $\{\mathscr{F}_{t\geq t_0>0}\}$ is its filtration on a complete filtered probability space $(\Omega, \mathscr{F}, \{\mathscr{F}_{t\geq t_0>0}\}, \mathscr{P})$ over $[t_0,t_f]$ with the correlation of independent increments

$$E\left\{[w(\tau_1) - w(\tau_2)][w(\tau_1) - w(\tau_2)]^T\right\} = W|\tau_1 - \tau_2|\,, \quad W > 0, \forall \tau_1, \tau_2 \in [t_0, t_f]\,.$$

The admissible control $u \in L^2_{\mathscr{F}_t}(\Omega;\mathscr{C}([t_0,t_f];\mathbb{R}^m))$ defined by a subset of Hilbert space of $\mathbb{R}^m$-valued square-integrable processes on $[t_0,t_f]$ that are adapted to the σ-field $\mathscr{F}_t$ generated by $w(t)$ with $E\left\{\int_{t_0}^{t_f} u^T(\tau)u(\tau)d\tau\right\} < \infty$ is robust against the stochastic elements ω so that the resulting system outputs $z \in L^2_{\mathscr{F}_t}(\Omega;\mathscr{C}([t_0,t_f];\mathbb{R}^r))$ effectively follow the desired transient responses $z_d \in L^2([t_0,t_f];\mathbb{R}^r)$ of the given linear reference model

$$dx_d(t) = A_d(t)x_d(t)dt\,, \quad x_d(t_0) = x_{d0}, \tag{4.3}$$

$$z_d(t) = C_d(t)x_d(t), \tag{4.4}$$

whereby the deterministic coefficients $A_d \in \mathscr{C}([t_0,t_f];\mathbb{R}^{n_d \times n_d})$ and $C_d \in \mathscr{C}([t_0,t_f];\mathbb{R}^{r \times n_d})$.

In addition, the stochastic system (4.1) in the absence of process noises is supposed to be uniformly exponentially stable. That is, there exist positive constants η_1 and η_2 such that the pointwise matrix norm of the closed-loop state transition matrix satisfies the inequality

$$||\Phi(t,\tau)|| \leq \eta_1 e^{-\eta_2(t-\tau)} \qquad \forall\, t \geq \tau \geq t_0.$$

The pair $(A(t),B(t))$ is stabilizable if there exists a bounded matrix-valued function $K(t)$ such that $dx(t) = (A(t)+B(t)K(t))x(t)dt$ is uniformly exponentially stable. Similarly, the pair $(C_d(t),A_d(t))$ is assumed detectable. Hence, there must exist a bounded matrix-valued function $L_d(t)$ so that $dx_d(t) = (A_d(t) - L_d(t)C_d(t))x_d(t)dt$ is also uniformly exponentially stable.

Associated with the admissible $(x(\cdot),x_{d0}(\cdot);u(\cdot))$ is performance appraisal of a finite-horizon integral-quadratic-form cost $J : \mathbb{R}^n \times \mathbb{R}^{n_d} \times L^2_{\mathscr{F}_t}(\Omega;\mathscr{C}([t_0,t_f];\mathbb{R}^m)) \mapsto \mathbb{R}^+$

$$\begin{aligned} J(x_0,x_{d0};u(\cdot)) &= (z(t_f)-z_d(t_f))^T Q_f (z(t_f)-z_d(t_f)) \\ &\quad + \int_{t_0}^{t_f} [(z(\tau)-z_d(\tau))^T Q(\tau)(z(\tau)-z_d(\tau)) + u^T(\tau)R(\tau)u(\tau)]d\tau, \end{aligned} \tag{4.5}$$

where the design parameters $Q_f \in \mathbb{R}^{r \times r}$, $Q \in \mathscr{C}([t_0,t_f];\mathbb{R}^{r \times r})$ and invertible $R \in \mathscr{C}([t_0,t_f];\mathbb{R}^{m \times m})$ are deterministic, symmetric, and positive semidefinite relative weighting of the terminal state, state trajectory, and control input.

In view of the linear system (4.1)–(4.2) and the quadratic performance measure (4.5), admissible control laws are therefore, restricted to the linear mapping $\gamma : [t_0,t_f] \times L^2_{\mathscr{F}_t}(\Omega;\mathscr{C}([t_0,t_f];\mathbb{R}^n)) \times L^2(\mathscr{C}([t_0,t_f];\mathbb{R}^{n_d})) \mapsto L^2_{\mathscr{F}_t}(\Omega;\mathscr{C}([t_0,t_f];\mathbb{R}^m))$ with the rule of actions

$$u(t) = \gamma(t,x(t),x_d(t)) \triangleq K_x(t)x(t) + K_{x_d}(t)x_d(t), \tag{4.6}$$

where the deterministic matrix-valued functions $K_x \in \mathscr{C}([t_0,t_f];\mathbb{R}^{m \times n})$ and $K_{x_d} \in \mathscr{C}([t_0,t_f];\mathbb{R}^{m \times n_d})$ are, respectively, the admissible feedback gain and the feedforward gain on the reference model states from restraint sets which are yet to be defined. To convert the stochastic servo problem to a stochastic regulator problem, it requires to define augmented state variables and system parameters

$$x_a \triangleq \begin{bmatrix} x \\ x_d \end{bmatrix}; \quad x_{a0} = \begin{bmatrix} x_0 \\ x_{d0} \end{bmatrix}; \quad A_a \triangleq \begin{bmatrix} A+BK_x & BK_{x_d} \\ 0 & A_d \end{bmatrix}; \quad G_a \triangleq \begin{bmatrix} G \\ 0 \end{bmatrix}$$

together with the state and terminal penalty weightings

$$N_a \triangleq \begin{bmatrix} C^T QC + K_x^T RK_x & -C^T QC_\mathrm{d} + K_x^T RK_{x_d} \\ -C_\mathrm{d}^T QC + K_{x_d}^T RK_x & C_\mathrm{d}^T QC_\mathrm{d} + K_{x_d}^T RK_{x_d} \end{bmatrix};$$

$$Q_{fa} \triangleq \begin{bmatrix} C^T(t_f)Q_f C(t_f) & -C^T(t_f)Q_f C_\mathrm{d}(t_f) \\ -C_\mathrm{d}^T(t_f)Q_f C(t_f) & C_\mathrm{d}^T(t_f)Q_f C_\mathrm{d}(t_f) \end{bmatrix},$$

which thus lead to the equivalent performance measure of Eq. (4.5)

$$J(x_{a0}; K_x(\cdot), K_{x_d}(\cdot)) = x_a^T(t_f)Q_{fa}(t_f)x_a(t_f) + \int_{t_0}^{t_f} x_a^T(\tau)N_a(\tau)x_a(\tau)\mathrm{d}\tau \tag{4.7}$$

subject to the augmented system of Eqs. (4.1)–(4.4)

$$\mathrm{d}x_a(t) = A_a(t)x_a(t)\mathrm{d}t + G_a(t)\mathrm{d}w(t), \quad x_a(t_0) = x_{a0}. \tag{4.8}$$

As previously established by Chaps. 2 and 3, the following result contains an efficient and tractable procedure for calculating all the performance-measure statistics of any order that completely capture the performance uncertainty of the augmented stochastic system (4.7) and (4.8).

Theorem 4.2.1 (Performance-Measure Statistics). *Suppose that (A,B) is uniformly stabilizable and $(C_\mathrm{d},A_\mathrm{d})$ is uniformly detectable. The kth-order statistic of the chi-squared performance measure (4.7) is given by*

$$\kappa_k = x_{a0}^T H_a(t_0,k)x_{a0} + D_a(t_0,k), \quad k \in \mathbb{Z}^+, \tag{4.9}$$

whereby the cumulant-generating components $H_a(\tau,k)$ and $D_a(\tau,k)$ evaluated at $\tau = t_0$ satisfy the cumulant-generating equations (with the dependence of $H_a(\tau,k)$ and $D_a(\tau,k)$ upon K_x and K_{x_d} suppressed)

$$\frac{\mathrm{d}}{\mathrm{d}\tau}H_a(\tau,1) = -A_a^T(\tau)H_a(\tau,1) - H_a(\tau,1)A_a(\tau) - N_a(\tau), \tag{4.10}$$

$$\frac{\mathrm{d}}{\mathrm{d}\tau}H_a(\tau,r) = -A_a^T(\tau)H_a(\tau,r) - H_a(\tau,r)A_a(\tau) - \sum_{s=1}^{r-1}\frac{2r!}{s!(r-s)!}H_a(\tau,s)G_a(\tau)WG_a^T(\tau)H_a(\tau,r-s), \quad 2 \le r \le k, \tag{4.11}$$

$$\frac{\mathrm{d}}{\mathrm{d}\tau}D_a(\tau,r) = -\mathrm{Tr}\left\{H_a(\tau,r)G_a(\tau)WG_a^T(\tau)\right\}, \quad 1 \le r \le k \tag{4.12}$$

with the terminal-value conditions $H_a(t_f,1) = Q_{fa}(t_f)$, $H_a(t_f,r) = 0$ for $2 \le r \le k$, and $D_a(t_f,r) = 0$ for $1 \le r \le k$.

Roughly speaking, this computational procedure now allows the incorporation of classes of linear feedback strategies in the statistical optimal control problems. Moreover, these performance-measure statistics or cost cumulants (4.9) are further interpreted in terms of variables and system parameters of the original servo problem

$$\kappa_k = x_0^T H_{11}^k(t_0)x_0 + 2x_0^T H_{12}^k(t_0)z_0 + z_0^T H_{22}^k(t_0)z_0 + D^k(t_0)\,, \quad k \in \mathbb{Z}^+ \tag{4.13}$$

provided that the matrix partition is of the form

$$H_a(\tau,r) = \begin{bmatrix} H_{11}^r(\tau) & H_{12}^r(\tau) \\ (H_{12}^r)^T(\tau) & H_{22}^r(\tau) \end{bmatrix}, \quad 1 \le r \le k \tag{4.14}$$

from which the components $\{H_{11}^r(\tau)\}_{r=1}^k$, $\{H_{12}^r(\tau)\}_{r=1}^k$, and $\{H_{22}^r(\tau)\}_{r=1}^k$ implicitly depend on K_x and K_{x_d} and satisfy the backward-in-time matrix-valued differential equations

$$\begin{aligned}\frac{d}{d\tau}H_{11}^1(\tau) &= -[A(\tau)+B(\tau)K_x(\tau)]^T H_{11}^1(\tau) - H_{11}^1(\tau)[A(\tau)+B(\tau)K_x(\tau)] \\ &\quad - K_x^T(\tau)R(\tau)K_x(\tau) - C^T(\tau)Q(\tau)C(\tau),\end{aligned} \tag{4.15}$$

$$\begin{aligned}\frac{d}{d\tau}H_{11}^r(\tau) &= -[A(\tau)+B(\tau)K_x(\tau)]^T H_{11}^r(\tau) - H_{11}^r(\tau)[A(\tau)+B(\tau)K_x(\tau)] \\ &\quad - \sum_{s=1}^{r-1} \frac{2r!}{s!(r-s)!} H_{11}^s(\tau)G(\tau)WG^T(\tau)H_{11}^{r-s}(\tau), \; 2 \le r \le k,\end{aligned} \tag{4.16}$$

$$\begin{aligned}\frac{d}{d\tau}H_{12}^1(\tau) &= -[A(\tau)+B(\tau)K_x(\tau)]^T H_{12}^1(\tau) - H_{11}^1(\tau)B(\tau)K_{x_d}(\tau) - H_{12}^1(\tau)A_d(\tau) \\ &\quad - K_x^T(\tau)R(\tau)K_{x_d}(\tau) + C^T(\tau)Q(\tau)C_d(\tau),\end{aligned} \tag{4.17}$$

$$\begin{aligned}\frac{d}{d\tau}H_{12}^r(\tau) &= -[A(\tau)+B(\tau)K_x(\tau)]^T H_{12}^r(\tau) - H_{11}^r(\tau)B(\tau)K_{x_d}(\tau) - H_{12}^r(\tau)A_d(\tau) \\ &\quad - \sum_{s=1}^{r-1} \frac{2r!}{s!(r-s)!} H_{11}^s(\tau)G(\tau)WG^T(\tau)H_{12}^{r-s}(\tau), \quad 2 \le r \le k,\end{aligned} \tag{4.18}$$

$$\begin{aligned}\frac{d}{d\tau}H_{22}^1(\tau) &= -A_d^T(\tau)H_{22}^1(\tau) - H_{22}^1(\tau)A_d(\tau) - K_{x_d}^T(\tau)B^T(\tau)H_{12}^1(\tau) \\ &\quad - (H_{12}^1)^T(\tau)B(\tau)K_{x_d}(\tau) - K_{x_d}^T(\tau)R(\tau)K_{x_d}(\tau) - C_d^T(\tau)Q(\tau)C_d(\tau),\end{aligned} \tag{4.19}$$

$$\begin{aligned}\frac{d}{d\tau}H_{22}^r(\tau) &= -A_d^T(\tau)H_{22}^r(\tau) - H_{22}^r(\tau)A_d(\tau) - K_{x_d}^T(\tau)B^T(\tau)H_{12}^r(\tau) \\ &\quad - (H_{12}^r)^T(\tau)B(\tau)K_{x_d}(\tau) - \sum_{s=1}^{r-1} \frac{2r!}{s!(r-s)!} (H_{12}^s)^T(\tau)G(\tau)WG^T(\tau)H_{12}^{r-s}(\tau),\end{aligned} \tag{4.20}$$

$$\frac{d}{d\tau}D^r(\tau) = -\mathrm{Tr}\{H_{11}^r(\tau)G(\tau)WG^T(\tau)\} \tag{4.21}$$

with terminal-value conditions $H_{11}^1(t_f) = C^T(t_f)Q_f C(t_f)$, $H_{11}^r(t_f) = 0$ for $2 \le r \le k$; $H_{12}^1(t_f) = -C^T(t_f)Q_f C_{\mathrm{d}}(t_f)$, $H_{12}^r(t_f) = 0$ for $2 \le r \le k$; $H_{22}^1(t_f) = C_{\mathrm{d}}^T(t_f)Q_f C_{\mathrm{d}}(t_f)$, $H_{22}^r(t_f) = 0$ for $2 \le r \le k$ and $D^r(t_f) = 0$ for $1 \le r \le k$.

4.3 The System Control Problem

In an attempt to describe or model performance uncertainty, the essence of information about these higher-order performance-measure statistics of Eq. (4.7) is now served as a source of information flow, which will affect perception of the problem and the environment at the design of feedback controller with risk consequences. This consideration therefore makes statistical optimal control problems herein particularly unique as compared with the more traditional dynamic programming class of investigations. In other words, the time-backward trajectories (4.15)–(4.21) should be considered as the "new" dynamics by which the resulting dynamic optimization of Mayer type [8] therefore depends on these "new" state variables $H_{11}^r(\tau)$, $H_{12}^r(\tau)$, $H_{22}^r(\tau)$ and $D^r(\tau)$, not the classical states $x(t)$ as the people may often expect.

For notational simplicity and compact formulation, it is required to introduce the convenient mappings to denote the right members of Eqs. (4.15)–(4.21)

$$\begin{aligned}
\mathscr{F}_{11}^r &: [t_0,t_f] \times (\mathbb{R}^{n\times n})^k \times \mathbb{R}^{m\times n} \mapsto \mathbb{R}^{n\times n}, \\
\mathscr{F}_{12}^r &: [t_0,t_f] \times (\mathbb{R}^{n\times n_d})^k \times \mathbb{R}^{m\times n} \times \mathbb{R}^{m\times n_d} \mapsto \mathbb{R}^{n\times n_d}, \\
\mathscr{F}_{22}^r &: [t_0,t_f] \times (\mathbb{R}^{n_d\times n_d})^k \times (\mathbb{R}^{n\times n_d})^k \times \mathbb{R}^{m\times n_d} \mapsto \mathbb{R}^{n_d\times n_d}, \\
\mathscr{G}^r &: [t_0,t_f] \times (\mathbb{R}^{n})^k \mapsto \mathbb{R}
\end{aligned}$$

with the rules of action

$$\begin{aligned}
\mathscr{F}_{11}^1(\tau,\mathscr{H}_{11},K_x) &\triangleq -[A(\tau)+B(\tau)K_x(\tau)]^T\,\mathscr{H}_{11}^1(\tau) - \mathscr{H}_{11}^1(\tau)\,[A(\tau)+B(\tau)K_x(\tau)] \\
&\quad - K_x^T(\tau)R(\tau)K_x(\tau) - C^T(\tau)Q(\tau)C(\tau), \\
\mathscr{F}_{11}^r(\tau,\mathscr{H}_{11},K_x) &\triangleq -[A(\tau)+B(\tau)K_x(\tau)]^T\,\mathscr{H}_{11}^r(\tau) - \mathscr{H}_{11}^r(\tau)\,[A(\tau)+B(\tau)K_x(\tau)] \\
&\quad - \sum_{s=1}^{r-1}\frac{2r!}{s!(r-s)!}\mathscr{H}_{11}^s(\tau)G(\tau)WG^T(\tau)\mathscr{H}_{11}^{r-s}(\tau), \\
\mathscr{F}_{12}^1(\tau,\mathscr{H}_{12},K_x,K_{x_d}) &\triangleq -[A(\tau)+B(\tau)K_x(\tau)]^T\,\mathscr{H}_{12}^1(\tau) - \mathscr{H}_{11}^1(\tau)B(\tau)K_{x_d}(\tau) \\
&\quad - \mathscr{H}_{12}^1(\tau)A_{\mathrm{d}}(\tau) - K_x^T(\tau)R(\tau)K_{x_d}(\tau) + C^T(\tau)Q(\tau)C_{\mathrm{d}}(\tau), \\
\mathscr{F}_{12}^r(\tau,\mathscr{H}_{12},K_x,K_{x_d}) &\triangleq -[A(\tau)+B(\tau)K_x(\tau)]^T\,\mathscr{H}_{12}^r(\tau) - \mathscr{H}_{11}^r(\tau)B(\tau)K_{x_d}(\tau) \\
&\quad - \mathscr{H}_{12}^r(\tau)A_{\mathrm{d}}(\tau) - \sum_{s=1}^{r-1}\frac{2r!}{s!(r-s)!}\mathscr{H}_{11}^s(\tau)G(\tau)WG^T(\tau)\mathscr{H}_{12}^{r-s}(\tau),
\end{aligned}$$

$$\mathscr{F}_{22}^1(\tau,\mathscr{H}_{22},\mathscr{H}_{12},K_{x_d}) \triangleq -A_d^T(\tau)\mathscr{H}_{22}^1(\tau) - \mathscr{H}_{22}^1(\tau)A_d(\tau)$$
$$- K_{x_d}^T(\tau)B^T(\tau)\mathscr{H}_{12}^1(\tau) - (\mathscr{H}_{12}^1)^T(\tau)B(\tau)K_{x_d}(\tau)$$
$$- K_{x_d}^T(\tau)R(\tau)K_{x_d}(\tau) - C_d^T(\tau)Q(\tau)C_d(\tau),$$
$$\mathscr{F}_{22}^r(\tau,\mathscr{H}_{22},\mathscr{H}_{12},K_{x_d}) \triangleq -A_d^T(\tau)\mathscr{H}_{22}^r(\tau) - \mathscr{H}_{22}^r(\tau)A_d(\tau)$$
$$- K_{x_d}^T(\tau)B^T(\tau)\mathscr{H}_{12}^r(\tau) - (\mathscr{H}_{12}^r)^T(\tau)B(\tau)K_{x_d}(\tau)$$
$$- \sum_{s=1}^{r-1}\frac{2r!}{s!(r-s)!}(\mathscr{H}_{12}^s)^T(\tau)G(\tau)WG^T(\tau)\mathscr{H}_{12}^{r-s}(\tau),$$

$$\mathscr{G}^r(\tau,\mathscr{H}_{11}) \triangleq -\mathrm{Tr}\{\mathscr{H}_{11}^r(\tau)G(\tau)WG^T(\tau)\},$$

where the components of k-tuple variables $\mathscr{H}_{11}$, $\mathscr{H}_{12}$, $\mathscr{H}_{22}$ and $\mathscr{D}$ defined by

$$\mathscr{H}_{11}(\cdot) \triangleq (\mathscr{H}_{11}^1(\cdot),\dots,\mathscr{H}_{11}^k(\cdot)), \qquad \mathscr{H}_{12}(\cdot) \triangleq (\mathscr{H}_{12}^1(\cdot),\dots,\mathscr{H}_{12}^k(\cdot)),$$
$$\mathscr{H}_{22}(\cdot) \triangleq (\mathscr{H}_{22}^1(\cdot),\dots,\mathscr{H}_{22}^k(\cdot)), \qquad \mathscr{D}(\cdot) \triangleq (\mathscr{D}^1(\cdot),\dots,\mathscr{D}^k(\cdot))$$

provided that each element $\mathscr{H}_{11}^r \in \mathscr{C}^1([t_0,t_f];\mathbb{R}^{n\times n})$, $\mathscr{H}_{12}^r \in \mathscr{C}^1([t_0,t_f];\mathbb{R}^{n\times n_d})$, $\mathscr{H}_{22}^r \in \mathscr{C}^1([t_0,t_f];\mathbb{R}^{n_d\times n_d})$ and $\mathscr{D}^r \in \mathscr{C}^1([t_0,t_f];\mathbb{R})$ have the representations $\mathscr{H}_{11}^r(\cdot) \triangleq H_{11}^r(\cdot)$, $\mathscr{H}_{12}^r(\cdot) \triangleq H_{12}^r(\cdot)$, $\mathscr{H}_{22}^r(\cdot) \triangleq H_{22}^r(\cdot)$ and $\mathscr{D}^r(\cdot) \triangleq D^r(\cdot)$.

Further on, the product mappings that are bounded and Lipschitz continuous on $[t_0,t_f]$ of the dynamical equations (4.15)–(4.21)

$$\mathscr{F}_{11} : [t_0,t_f] \times (\mathbb{R}^{n\times n})^k \times \mathbb{R}^{m\times n} \mapsto (\mathbb{R}^{n\times n})^k,$$
$$\mathscr{F}_{12} : [t_0,t_f] \times (\mathbb{R}^{n\times n_d})^k \times \mathbb{R}^{m\times n} \times \mathbb{R}^{m\times n_d} \mapsto (\mathbb{R}^{n\times n_d})^k,$$
$$\mathscr{F}_{22} : [t_0,t_f] \times (\mathbb{R}^{n_d\times n_d})^k \times (\mathbb{R}^{n\times n_d})^k \times \mathbb{R}^{m\times n_d} \mapsto (\mathbb{R}^{n_d\times n_d})^k,$$
$$\mathscr{G} : [t_0,t_f] \times (\mathbb{R}^{n\times n})^k \mapsto \mathbb{R}^k$$

in the statistical optimal control for servo problems have the rules of action given by

$$\frac{d}{d\tau}\mathscr{H}_{11}(\tau) = \mathscr{F}_{11}(\tau,\mathscr{H}_{11}(\tau),K_x(\tau)), \quad \mathscr{H}_{11}(t_f) = \mathscr{H}_{11}^f, \tag{4.22}$$

$$\frac{d}{d\tau}\mathscr{H}_{12}(\tau) = \mathscr{F}_{12}(\tau,\mathscr{H}_{12}(\tau),K_x(\tau),K_{x_d}(\tau)), \quad \mathscr{H}_{12}(t_f) = \mathscr{H}_{12}^f, \tag{4.23}$$

$$\frac{d}{d\tau}\mathscr{H}_{22}(\tau) = \mathscr{F}_{22}(\tau,\mathscr{H}_{22}(\tau),\mathscr{H}_{12}(\tau),K_x(\tau)), \quad \mathscr{H}_{22}(t_f) = \mathscr{H}_{22}^f, \tag{4.24}$$

$$\frac{d}{d\tau}\mathscr{D}(\tau) = G(\tau,\mathscr{H}_{11}(\tau)), \quad \mathscr{D}(t_f) = \mathscr{D}_f \tag{4.25}$$

under the essential definitions

$$\mathscr{F}_{11} \triangleq \mathscr{F}_{11}^1 \times \cdots \times \mathscr{F}_{11}^k, \qquad \mathscr{F}_{12} \triangleq \mathscr{F}_{12}^1 \times \cdots \times \mathscr{F}_{12}^k,$$
$$\mathscr{F}_{22} \triangleq \mathscr{F}_{22}^1 \times \cdots \times \mathscr{F}_{22}^k, \qquad \mathscr{G} \triangleq \mathscr{G}^1 \times \cdots \times \mathscr{G}^k$$

and the terminal-value conditions

$$\mathscr{H}_{11}^f = (C^T(t_f)Q_fC(t_f), 0, \ldots, 0), \qquad \mathscr{H}_{12}^f = (-C^T(t_f)Q_fC_\mathrm{d}(t_f), 0, \ldots, 0),$$
$$\mathscr{H}_{22}^f = (C_\mathrm{d}^T(t_f)Q_fC_\mathrm{d}(t_f), 0, \ldots, 0), \qquad \mathscr{D}^f = (0, \ldots, 0).$$

Recall that the product system uniquely determines $\mathscr{H}_{11}$, $\mathscr{H}_{12}$, $\mathscr{H}_{22}$, and $\mathscr{D}$ once admissible feedback and feedforward gains K_x and K_{x_d} are specified. Therefore, it should be considered $\mathscr{H}_{11} \equiv \mathscr{H}_{11}(\cdot, K_x, K_{x_d})$, $\mathscr{H}_{12} \equiv \mathscr{H}_{12}(\cdot, K_x, K_{x_d})$, $\mathscr{H}_{22} \equiv \mathscr{H}_{22}(\cdot, K_x, K_{x_d})$, and $\mathscr{D} \equiv \mathscr{D}(\cdot, K_x, K_{x_d})$.

When measuring performance reliability, statistical analysis for probabilistic nature of performance uncertainty is relied on as part of the long range assessment of reliability. One of the most widely used measures for performance reliability is the statistical mean or average to summarize the underlying performance variations. However, other aspects of performance distributions, that do not appear in most of the existing progress, are variance, skewness, flatness, etc. With that said, the performance index with risk-value awareness in the statistical control problem can now be formulated in K_x and K_{x_d}.

Definition 4.3.1 (Risk-Value Aware Performance Index). Fix $k \in \mathbb{Z}^+$ and the sequence $\mu = \{\mu_i \geq 0\}_{i=1}^k$ with $\mu_1 > 0$. Then for the given 3-tuple (t_0, x_0, x_{d0}), the performance index with risk consequences for the stochastic servo systems is given by

$$\phi_0^s : \{t_0\} \times (\mathbb{R}^{n \times n})^k \times (\mathbb{R}^{n \times n_d})^k \times (\mathbb{R}^{n_d \times n_d})^k \times \mathbb{R}^k \mapsto \mathbb{R}^+$$

with the rule of action

$$\phi_0^s(t_0, \mathscr{H}_{11}(t_0), \mathscr{H}_{12}(t_0), \mathscr{H}_{22}(t_0), \mathscr{D}(t_0)) \triangleq \underbrace{\mu_1 \kappa_1}_{\text{Value Measure}} + \underbrace{\mu_2 \kappa_2 + \cdots + \mu_k \kappa_k}_{\text{Risk Measures}}$$
$$= \sum_{r=1}^k \mu_r \left[x_0^T \mathscr{H}_{11}^r(t_0) x_0 + 2x_0^T \mathscr{H}_{12}^r(t_0) x_{d0} + x_{d0}^T \mathscr{H}_{22}^r(t_0) x_{d0} + \mathscr{D}^r(t_0)\right], \quad (4.26)$$

where additional parametric design of freedom μ_r, chosen by risk-averse controller designers, represents different levels of robustness prioritization according to the importance of the resulting performance-measure statistics to the probabilistic performance distribution. And the unique solutions $\{\mathscr{H}_{11}^r(\tau)\}_{r=1}^k$, $\{\mathscr{H}_{12}^r(\tau)\}_{r=1}^k$, $\{\mathscr{H}_{22}^r(\tau)\}_{r=1}^k$, and $\{\mathscr{D}^r(\tau)\}_{r=1}^k$ evaluated at $\tau = t_0$ satisfy the dynamical equations (4.22)–(4.25).

It is worth to observe the performance index (4.26) adopts a new and comprehensive optimization criterion which introduces additional parametric design of freedom in the class of feedback control laws that will then result in a broad class of problem solutions as one can directly derive from these solutions to other related results in LQG and risk-sensitive control problems. More importantly, the ultimate objective here is to introduce a competition among performance-measure statistics as they directly influence on the performance distribution of Eq. (4.5).

Definition 4.3.2 (Admissible Feedback and Feedforward Gains). For given terminal data $(t_f, \mathscr{H}_{11}^f, \mathscr{H}_{12}^f, \mathscr{H}_{22}^f, \mathscr{D}^f)$, the classes of admissible feedback and feedforward gains are defined as follows. Let compact subsets $\overline{K}_x \subset \mathbb{R}^{m\times n}$ and $\overline{K}_{x_d} \subset \mathbb{R}^{m\times n_d}$ be the sets of allowable gain values. With $k \in \mathbb{Z}^+$ and sequence $\mu = \{\mu_r \ge 0\}_{r=1}^k$ and $\mu_1 > 0$ given, the sets of admissible $\mathscr{K}^x_{t_f,\mathscr{H}_{11}^f,\mathscr{H}_{12}^f,\mathscr{H}_{22}^f,\mathscr{D}^f;\mu}$ and $\mathscr{K}^{x_d}_{t_f,\mathscr{H}_{11}^f,\mathscr{H}_{12}^f,\mathscr{H}_{22}^f,\mathscr{D}^f;\mu}$ are the classes of $\mathscr{C}([t_0,t_f];\mathbb{R}^{m\times n})$ and $\mathscr{C}([t_0,t_f];\mathbb{R}^{m\times n_d})$ with values $K_x(\cdot) \in \overline{K}_x$ and $K_{x_d}(\cdot) \in \overline{K}_{x_d}$ for which solutions to the dynamical equations (4.22)–(4.25) exist on the finite interval of optimization $[t_0,t_f]$.

The development in the sequel is motivated by the excellent treatment in [8] and is intended to follow it closely. Because the development therein embodies the traditional end-point problem and corresponding use of dynamic programming, it is necessary to make appropriate modifications in the sequence of results, as well as to introduce the terminology of statistical optimal control.

Definition 4.3.3 (Optimization Problem of Mayer Type). Suppose that $k \in \mathbb{Z}^+$ and the sequence $\mu = \{\mu_r \ge 0\}_{r=1}^k$ with $\mu_1 > 0$ are fixed. Then, the optimization problem over $[t_0,t_f]$ is defined as the minimization of the performance index (4.26) of Mayer type with respect to (K_x, K_{x_d}) in $\mathscr{K}^x_{t_f,\mathscr{H}_{11}^f,\mathscr{H}_{12}^f,\mathscr{H}_{22}^f,\mathscr{D}^f;\mu} \times \mathscr{K}^{x_d}_{t_f,\mathscr{H}_{11}^f,\mathscr{H}_{12}^f,\mathscr{H}_{22}^f,\mathscr{D}^f;\mu}$ and subject to the dynamical equations (4.22)–(4.25).

To embed the aforementioned optimization into a larger optimal control problem, the terminal time and states $(t_f, \mathscr{H}_{11}^f, \mathscr{H}_{12}^f, \mathscr{H}_{22}^f, \mathscr{D}^f)$ are parameterized as $(\varepsilon, \mathscr{Y}_{11}, \mathscr{Y}_{12}, \mathscr{Y}_{22}, \mathscr{Z})$ through the help of a reachable set.

Definition 4.3.4 (Reachable Set). Let the reachable set $\mathscr{Q} \triangleq \{(\varepsilon, \mathscr{Y}_{11}, \mathscr{Y}_{12}, \mathscr{Y}_{22}, \mathscr{Z}) \in [t_0,t_f] \times (\mathbb{R}^{n\times n})^k \times (\mathbb{R}^{n\times n_d})^k \times (\mathbb{R}^{n_d\times n_d})^k \times \mathbb{R}^k$ such that $\mathscr{K}^x_{t_f,\mathscr{H}_{11}^f,\mathscr{H}_{12}^f,\mathscr{H}_{22}^f,\mathscr{D}^f;\mu} \times \mathscr{K}^{x_d}_{t_f,\mathscr{H}_{11}^f,\mathscr{H}_{12}^f,\mathscr{H}_{22}^f,\mathscr{D}^f;\mu} \neq \emptyset\}$.

Therefore, the value function for this optimization problem is now depending on parameterizations of the terminal-value conditions.

Definition 4.3.5 (Value Function). Suppose $(\varepsilon, \mathscr{Y}_{11}, \mathscr{Y}_{12}, \mathscr{Y}_{22}, \mathscr{Z}) \in [t_0, t_f] \times (\mathbb{R}^{n\times n})^k \times (\mathbb{R}^{n\times n_d})^k \times (\mathbb{R}^{n_d\times n_d})^k \times \mathbb{R}^k$ be given. Then, the value function $\mathscr{V}(\varepsilon, \mathscr{Y}_{11}, \mathscr{Y}_{12}, \mathscr{Y}_{22}, \mathscr{Z})$ is defined by

$$\mathscr{V}(\varepsilon, \mathscr{Y}_{11}, \mathscr{Y}_{12}, \mathscr{Y}_{22}, \mathscr{Z}) = \inf_{K_x \in \overline{K}_x, K_{x_d} \in \overline{K}_z} \phi_0^s(t_0, \mathscr{H}_{11}(t_0), \mathscr{H}_{12}(t_0), \mathscr{H}_{22}(t_0), \mathscr{D}(t_0)).$$

It is conventional to let $\mathscr{V}(\varepsilon, \mathscr{Y}_{11}, \mathscr{Y}_{12}, \mathscr{Y}_{22}, \mathscr{Z}) = +\infty$ when $\mathscr{K}^x_{t_f, \mathscr{H}^f_{11}, \mathscr{H}^f_{12}, \mathscr{H}^f_{22}, \mathscr{D}^f; \mu} \times \mathscr{K}^{x_d}_{t_f, \mathscr{H}^f_{11}, \mathscr{H}^f_{12}, \mathscr{H}^f_{22}, \mathscr{D}^f; \mu}$ is empty.

By adapting to the initial-cost problem and the terminologies present in statistical optimal control, the Hamilton-Jacobi-Bellman (HJB) equation satisfied by the value function is derived from the excellent treatment in [8].

Theorem 4.3.1 (HJB Equation for Mayer Problem). *Let $(\varepsilon, \mathscr{Y}_{11}, \mathscr{Y}_{12}, \mathscr{Y}_{22}, \mathscr{Z})$ be any interior point of the reachable set $\mathscr{Q}$ at which the value function $\mathscr{V}(\varepsilon, \mathscr{Y}_{11}, \mathscr{Y}_{12}, \mathscr{Y}_{22}, \mathscr{Z})$ is differentiable. If there exist optimal gains $K_x^* \in \mathscr{K}^x_{t_f, \mathscr{H}^f_{11}, \mathscr{H}^f_{12}, \mathscr{H}^f_{22}, \mathscr{D}^f; \mu}$ and $K_{x_d}^* \in \mathscr{K}^{x_d}_{t_f, \mathscr{H}^f_{11}, \mathscr{H}^f_{12}, \mathscr{H}^f_{22}, \mathscr{D}^f; \mu}$, then the partial differential equation of dynamic programming*

$$\begin{aligned}
0 = \min_{K_x \in \overline{K}_x, K_{x_d} \in \overline{K}_z} \Big\{ & \frac{\partial}{\partial \varepsilon} \mathscr{V}(\varepsilon, \mathscr{Y}_{11}, \mathscr{Y}_{12}, \mathscr{Y}_{22}, \mathscr{Z}) \\
& + \frac{\partial}{\partial \operatorname{vec}(\mathscr{Y}_{11})} \mathscr{V}(\varepsilon, \mathscr{Y}_{11}, \mathscr{Y}_{12}, \mathscr{Y}_{22}, \mathscr{Z}) \operatorname{vec}(\mathscr{F}_{11}(\varepsilon, \mathscr{Y}_{11}, K_x)) \\
& + \frac{\partial}{\partial \operatorname{vec}(\mathscr{Y}_{12})} \mathscr{V}(\varepsilon, \mathscr{Y}_{11}, \mathscr{Y}_{12}, \mathscr{Y}_{22}, \mathscr{Z}) \operatorname{vec}(\mathscr{F}_{12}(\varepsilon, \mathscr{Y}_{12}, K_x, K_{x_d})) \\
& + \frac{\partial}{\partial \operatorname{vec}(\mathscr{Y}_{22})} \mathscr{V}(\varepsilon, \mathscr{Y}_{11}, \mathscr{Y}_{12}, \mathscr{Y}_{22}, \mathscr{Z}) \operatorname{vec}(\mathscr{F}_{22}(\varepsilon, \mathscr{Y}_{22}, \mathscr{Y}_{12}, K_{x_d})) \\
& + \frac{\partial}{\partial \operatorname{vec}(\mathscr{Z})} \mathscr{V}(\varepsilon, \mathscr{Y}_{11}, \mathscr{Y}_{12}, \mathscr{Y}_{22}, \mathscr{Z}) \operatorname{vec}(\mathscr{G}(\varepsilon, \mathscr{Y}_{11})) \Big\}
\end{aligned} \tag{4.27}$$

is satisfied. The boundary condition of Eq. (4.27) is given by

$$\mathscr{V}(t_0, \mathscr{H}_{11}(t_0), \mathscr{H}_{12}(t_0), \mathscr{H}_{22}(t_0), \mathscr{D}(t_0)) = \phi_s(t_0, \mathscr{H}_{11}(t_0), \mathscr{H}_{12}(t_0), \mathscr{H}_{22}(t_0), \mathscr{D}(t_0)).$$

Proof. A rigorous proof of the necessary condition herein can be found in [9]. □

Theorem 4.3.2 (Verification Theorem). *Fix $k \in \mathbb{Z}^+$ and let $\mathscr{W}(\varepsilon, \mathscr{Y}_{11}, \mathscr{Y}_{12}, \mathscr{Y}_{22}, \mathscr{Z})$ be a continuously differentiable solution of Eq. (4.27) which satisfies the boundary condition*

$$\mathscr{W}(t_0, \mathscr{H}_{11}(t_0), \mathscr{H}_{12}(t_0), \mathscr{H}_{22}(t_0), \mathscr{D}(t_0)) = \phi_0^s(t_0, \mathscr{H}_{11}(t_0), \mathscr{H}_{12}(t_0), \mathscr{H}_{22}(t_0), \mathscr{D}(t_0)).$$

Let $(t_f, \mathcal{H}_{11}^f, \mathcal{H}_{12}^f, \mathcal{H}_{22}^f, \mathcal{D}^f)$ *be in* $\mathcal{Q}$; (K_x, K_{x_d}) *2-tuple in* $\mathcal{K}^x_{t_f, \mathcal{H}_{11}^f, \mathcal{H}_{12}^f, \mathcal{H}_{22}^f, \mathcal{D}^f; \mu} \times \mathcal{K}^{x_d}_{t_f, \mathcal{H}_{11}^f, \mathcal{H}_{12}^f, \mathcal{H}_{22}^f, \mathcal{D}^f; \mu}$; $\mathcal{H}_{11}$, $\mathcal{H}_{12}$, $\mathcal{H}_{22}$ *and* $\mathcal{D}$ *the solutions of Eqs. (4.22)–(4.25). Then,* $\mathcal{W}(\tau, \mathcal{H}_{11}(\tau), \mathcal{H}_{12}(\tau), \mathcal{H}_{22}(\tau), \mathcal{D}(\tau))$ *is a time-backward increasing function of* τ. *If* $(K_x^*, K_{x_d}^*)$ *is in* $\mathcal{K}^x_{t_f, \mathcal{H}_{11}^f, \mathcal{H}_{12}^f, \mathcal{H}_{22}^f, \mathcal{D}^f; \mu} \times \mathcal{K}^{x_d}_{t_f, \mathcal{H}_{11}^f, \mathcal{H}_{12}^f, \mathcal{H}_{22}^f, \mathcal{D}^f; \mu}$ *defined on* $[t_0, t_f]$ *with the corresponding solutions* $\mathcal{H}_{11}^*$, $\mathcal{H}_{12}^*$, $\mathcal{H}_{22}^*$, *and* $\mathcal{D}^*$ *of the dynamical equations (4.22)–(4.25) such that, for* $\tau \in [t_0, t_f]$

$$
\begin{aligned}
0 = & \frac{\partial}{\partial \varepsilon} \mathcal{W}(\tau, \mathcal{H}_{11}^*(\tau), \mathcal{H}_{12}^*(\tau), \mathcal{H}_{22}^*(\tau), \mathcal{D}^*(\tau)) \\
& + \frac{\partial}{\partial \operatorname{vec}(\mathcal{Y}_{11})} \mathcal{W}(\tau, \mathcal{H}_{11}^*(\tau), \mathcal{H}_{12}^*(\tau), \mathcal{H}_{22}^*(\tau), \mathcal{D}^*(\tau)) \operatorname{vec}(\mathcal{F}_{11}(\tau, \mathcal{H}_{11}^*(\tau), K_x^*(\tau))) \\
& + \frac{\partial}{\partial \operatorname{vec}(\mathcal{Y}_{12})} \mathcal{W}(\tau, \mathcal{H}_{11}^*(\tau), \mathcal{H}_{12}^*(\tau), \mathcal{H}_{22}^*(\tau), \mathcal{D}^*(\tau)) \\
& \times \operatorname{vec}(\mathcal{F}_{12}(\tau, \mathcal{H}_{12}^*(\tau), K_x^*(\tau), K_{x_d}^*(\tau))) \\
& + \frac{\partial}{\partial \operatorname{vec}(\mathcal{Y}_{22})} \mathcal{W}(\tau, \mathcal{H}_{11}^*(\tau), \mathcal{H}_{12}^*(\tau), \mathcal{H}_{22}^*(\tau), \mathcal{D}^*(\tau)) \\
& \times \operatorname{vec}(\mathcal{F}_{22}(\tau, \mathcal{H}_{22}^*(\tau), \mathcal{H}_{12}^*(\tau), K_{x_d}^*(\tau))) \\
& + \frac{\partial}{\partial \operatorname{vec}(\mathcal{Z})} \mathcal{W}(\tau, \mathcal{H}_{11}^*(\tau), \mathcal{H}_{12}^*(\tau), \mathcal{H}_{22}^*(\tau), \mathcal{D}^*(\tau)) \operatorname{vec}(\mathcal{G}(\tau, \mathcal{H}_{11}^*(\tau)))
\end{aligned} \tag{4.28}
$$

then K_x^* *and* $K_{x_d}^*$ *are optimal feedback and feedforward gains. Moreover,*

$$
\mathcal{W}(\varepsilon, \mathcal{Y}_{11}, \mathcal{Y}_{12}, \mathcal{Y}_{22}, \mathcal{Z}) = \mathcal{V}(\varepsilon, \mathcal{Y}_{11}, \mathcal{Y}_{12}, \mathcal{Y}_{22}, \mathcal{Z}) \tag{4.29}
$$

where $\mathcal{V}(\varepsilon, \mathcal{Y}_{11}, \mathcal{Y}_{12}, \mathcal{Y}_{22}, \mathcal{Z})$ *is the value function.*

Proof. It is already contained in the rigorous proof of the sufficiency with the essential conditions aforementioned from [9]. □

4.4 Statistical Optimal Control Solution

Interestingly enough, the dynamic optimization under investigation herein is in "Mayer form" and can therefore be solved by an adaptation of the Mayer-form verification theorem of dynamic programming given in [8]. In conformity of the dynamic programming framework, it is required to denote the terminal time and states of a family of optimization problems by $(\varepsilon, \mathcal{Y}_{11}, \mathcal{Y}_{12}, \mathcal{Y}_{22}, \mathcal{Z})$ rather than $(t_f, \mathcal{H}_{11}^f, \mathcal{H}_{12}^f, \mathcal{H}_{22}^f, \mathcal{D}^f)$. Subsequently, the value of the optimization problem depends on the terminal conditions. For instance, for any $\varepsilon \in [t_0, t_f]$, the states of the equations (4.22)–(4.25) are denoted by $\mathcal{H}_{11}(\varepsilon) = \mathcal{Y}_{11}$, $\mathcal{H}_{12}(\varepsilon) = \mathcal{Y}_{12}$, $\mathcal{H}_{22}(\varepsilon) = \mathcal{Y}_{22}$, and $\mathcal{D}(\varepsilon) = \mathcal{Z}$. Then, the quadratic-affine nature of Eq. (4.26) implies that a solution to the HJB equation (4.27) is of the form as follows.

Fix $k \in \mathbb{Z}^+$ and let $(\varepsilon, \mathscr{Y}_{11}, \mathscr{Y}_{12}, \mathscr{Y}_{22}, \mathscr{Z})$ be any interior point of the reachable set $\mathscr{Q}$ at which the real-valued function $\mathscr{W}(\varepsilon, \mathscr{Y}_{11}, \mathscr{Y}_{12}, \mathscr{Y}_{22}, \mathscr{Z})$ described by

$$\begin{aligned}\mathscr{W}(\varepsilon, \mathscr{Y}_{11}, \mathscr{Y}_{12}, \mathscr{Y}_{22}, \mathscr{Z}) &= x_0^T \sum_{r=1}^{k} \mu_r(\mathscr{Y}_{11}^r + \mathscr{E}_{11}^r(\varepsilon))x_0 + 2x_0^T \sum_{r=1}^{k} \mu_r(\mathscr{Y}_{12}^r + \mathscr{E}_{12}^r(\varepsilon))x_{d0} \\ &\quad + \sum_{r=1}^{k} \mu_r(\mathscr{Z}^r + \mathscr{T}^r(\varepsilon)) + x_{d0}^T \sum_{r=1}^{k} \mu_r(\mathscr{Y}_{22}^r + \mathscr{E}_{22}^r(\varepsilon))x_{d0}\end{aligned} \tag{4.30}$$

is differentiable. Notice that the time-parametric functions $\mathscr{E}_{11}^r \in \mathscr{C}^1([t_0,t_f];\mathbb{R}^{n\times n})$, $\mathscr{E}_{12}^r \in \mathscr{C}^1([t_0,t_f];\mathbb{R}^{n\times n_d})$, $\mathscr{E}_{22}^r \in \mathscr{C}^1([t_0,t_f];\mathbb{R}^{n_d\times n_d})$, and $\mathscr{T}^r \in \mathscr{C}^1([t_0,t_f];\mathbb{R})$ are yet to be determined. Next the derivative of $\mathscr{W}(\varepsilon, \mathscr{Y}_{11}, \mathscr{Y}_{12}, \mathscr{Y}_{22}, \mathscr{Z})$ with respect to ε can be shown as illustrated in [9]

$$\begin{aligned}\frac{d}{d\varepsilon}\mathscr{W}(\varepsilon, \mathscr{Y}_{11}, \mathscr{Y}_{12}, \mathscr{Y}_{22}, \mathscr{Z}) &= x_0^T \sum_{r=1}^{k} \mu_r(\mathscr{F}_{11}^r(\varepsilon, \mathscr{Y}_{11}, K_x) + \frac{d}{d\varepsilon}\mathscr{E}_{11}^r(\varepsilon))x_0 \\ &\quad + 2x_0^T \sum_{r=1}^{k} \mu_r(\mathscr{F}_{12}^r(\varepsilon, \mathscr{Y}_{12}, K_x, K_{x_d}) + \frac{d}{d\varepsilon}\mathscr{E}_{12}^r(\varepsilon))x_{d0} \\ &\quad + x_{d0}^T \sum_{r=1}^{k} \mu_r(\mathscr{F}_{22}^r(\varepsilon, \mathscr{Y}_{22}, \mathscr{Y}_{12}, K_{x_d}) + \frac{d}{d\varepsilon}\mathscr{E}_{22}^r(\varepsilon))x_{d0} \\ &\quad + \sum_{r=1}^{k} \mu_r(\mathscr{G}^r(\varepsilon, \mathscr{Y}_{11}) + \frac{d}{d\varepsilon}\mathscr{T}^r(\varepsilon))\end{aligned} \tag{4.31}$$

provided that the admissible 2-tuple $(K_x, K_{x_d}) \in \overline{K}_x \times \overline{K}_{x_d}$.

Replacing the candidate solution of Eq. (4.30) and the result (4.31) into the HJB equation (4.27), one obtains

$$\begin{aligned}\min_{(K_x,K_{x_d})\in\overline{K}_x\times\overline{K}_{x_d}} \Bigg\{ & x_0^T \sum_{r=1}^{k} \mu_r(\mathscr{F}_{11}^r(\varepsilon, \mathscr{Y}_{11}, K_x) + \frac{d}{d\varepsilon}\mathscr{E}_{11}^r(\varepsilon))x_0 \\ & + 2x_0^T \sum_{r=1}^{k} \mu_r(\mathscr{F}_{12}^r(\varepsilon, \mathscr{Y}_{12}, K_x, K_{x_d}) + \frac{d}{d\varepsilon}\mathscr{E}_{12}^r(\varepsilon))x_{d0} \\ & + x_{d0}^T \sum_{r=1}^{k} \mu_r(\mathscr{F}_{22}^r(\varepsilon, \mathscr{Y}_{22}, \mathscr{Y}_{12}, K_{x_d}) + \frac{d}{d\varepsilon}\mathscr{E}_{22}^r(\varepsilon))x_{d0} \\ & + \sum_{r=1}^{k} \mu_r(\mathscr{G}^r(\varepsilon, \mathscr{Y}_{11}) + \frac{d}{d\varepsilon}\mathscr{T}^r(\varepsilon)) \Bigg\} \equiv 0.\end{aligned} \tag{4.32}$$

Differentiating the expression within the bracket (4.32) with respect to K_x and K_{x_d} yields the necessary conditions for an interior extremum of the performance index with risk consequences (4.26) on $[t_0, t_f]$. In other words, the extremizing K_x and K_{x_d} must be

$$K_x(\varepsilon, \mathscr{Y}_{11}) = -R^{-1}(\varepsilon)B^T(\varepsilon)\sum_{r=1}^{k}\hat{\mu}_r\mathscr{Y}_{11}^r, \tag{4.33}$$

$$K_{x_d}(\varepsilon, \mathscr{Y}_{12}) = -R^{-1}(\varepsilon)B^T(\varepsilon)\sum_{r=1}^{k}\hat{\mu}_r\mathscr{Y}_{12}^r, \qquad \hat{\mu}_r = \mu_r/\mu_1\,. \tag{4.34}$$

In view of Eqs. (4.33) and (4.34), the value of the expression inside of the bracket of Eq. (4.32) becomes

$$x_0^T\Bigg[\sum_{r=1}^{k}\mu_r\frac{\mathrm{d}}{\mathrm{d}\varepsilon}\mathscr{E}_{11}^r(\varepsilon) - A^T(\varepsilon)\sum_{r=1}^{k}\mu_r\mathscr{Y}_{11}^r - \sum_{r=1}^{k}\mu_r\mathscr{Y}_{11}^rA(\varepsilon) - \mu_1C^T(\varepsilon)Q(\varepsilon)C(\varepsilon)$$
$$+\sum_{r=1}^{k}\hat{\mu}_r\mathscr{Y}_{11}^rB(\varepsilon)R^{-1}(\varepsilon)B^T(\varepsilon)\sum_{s=1}^{k}\mu_s\mathscr{Y}_{11}^s + \sum_{r=1}^{k}\mu_r\mathscr{Y}_{11}^rB(\varepsilon)R^{-1}(\varepsilon)B^T(\varepsilon)\sum_{s=1}^{k}\hat{\mu}_s\mathscr{Y}_{11}^r$$
$$-\mu_1\sum_{r=1}^{k}\hat{\mu}_r\mathscr{Y}_{11}^rB(\varepsilon)R^{-1}(\varepsilon)B^T(\varepsilon)\sum_{s=1}^{k}\hat{\mu}_s\mathscr{Y}_{11}^s\Bigg]x_0 + 2x_0^T\Bigg[\sum_{r=1}^{k}\mu_r\frac{\mathrm{d}}{\mathrm{d}\varepsilon}\mathscr{E}_{12}^r(\varepsilon)$$
$$-A^T(\varepsilon)\sum_{r=1}^{k}\mu_r\mathscr{Y}_{12}^r - \sum_{r=1}^{k}\mu_r\mathscr{Y}_{12}^rA_{\mathrm{d}}(\varepsilon) + \sum_{r=1}^{k}\hat{\mu}_r\mathscr{Y}_{11}^rB(\varepsilon)R^{-1}(\varepsilon)B^T(\varepsilon)\sum_{s=1}^{k}\mu_s\mathscr{Y}_{12}^s$$
$$+\sum_{r=1}^{k}\mu_r\mathscr{Y}_{11}^rB(\varepsilon)R^{-1}(\varepsilon)B^T(\varepsilon)\sum_{r=1}^{k}\hat{\mu}_r\mathscr{Y}_{12}^r + \mu_1C^T(\varepsilon)Q(\varepsilon)C_{\mathrm{d}}(\varepsilon)$$
$$-\mu_1\sum_{r=1}^{k}\hat{\mu}_r\mathscr{Y}_{11}^rB(\varepsilon)R^{-1}(\varepsilon)B^T(\varepsilon)\sum_{s=1}^{k}\hat{\mu}_s\mathscr{Y}_{12}^s$$
$$-\sum_{r=2}^{k}\mu_r\sum_{s=1}^{r-1}\frac{2r!}{s!(r-s)!}\mathscr{Y}_{11}^sG(\varepsilon)WG^T(\varepsilon)\mathscr{Y}_{12}^{r-s}\Bigg]x_{d0} + x_{d0}^T\Bigg[\sum_{r=1}^{k}\mu_r\frac{\mathrm{d}}{\mathrm{d}\varepsilon}\mathscr{E}_{22}^r(\varepsilon)$$
$$-A_{\mathrm{d}}^T(\varepsilon)\sum_{r=1}^{k}\mu_r\mathscr{Y}_{22}^r - \sum_{r=1}^{k}\mu_r\mathscr{Y}_{22}^rA_{\mathrm{d}}(\varepsilon) + \sum_{r=1}^{k}\hat{\mu}_r\mathscr{Y}_{12}^rB(\varepsilon)R^{-1}(\varepsilon)B^T(\varepsilon)\sum_{s=1}^{k}\mu_s\mathscr{Y}_{12}^s$$
$$+\sum_{r=1}^{k}\mu_r(\mathscr{Y}_{12}^r)^TB(\varepsilon)R^{-1}(\varepsilon)B^T(\varepsilon)\sum_{s=1}^{k}\hat{\mu}_s\mathscr{Y}_{12}^s - \mu_1C_{\mathrm{d}}^T(\varepsilon)Q(\varepsilon)C_{\mathrm{d}}(\varepsilon)$$
$$-\mu_1\sum_{r=1}^{k}\hat{\mu}_r\mathscr{Y}_{12}^rB(\varepsilon)R^{-1}(\varepsilon)B^T(\varepsilon)\sum_{s=1}^{k}\hat{\mu}_s\mathscr{Y}_{12}^s$$

$$
\left. - \sum_{r=2}^{k} \mu_r \sum_{s=1}^{r-1} \frac{2r!}{s!(r-s)!} (\mathscr{Y}_{12}^s)^T G(\varepsilon) W G^T(\varepsilon) \mathscr{Y}_{12}^{r-s} \right] x_{d0}
$$

$$
+ \sum_{r=1}^{k} \mu_r \frac{\mathrm{d}}{\mathrm{d}\varepsilon} \mathscr{T}^r(\varepsilon) - \sum_{r=1}^{k} \mu_r \mathrm{Tr}\left\{ \mathscr{Y}_{11}^r G(\varepsilon) W G^T(\varepsilon) \right\}. \tag{4.35}
$$

The minimum of Eq. (4.35) equal to zero for any $\varepsilon \in [t_0, t_f]$ when $\mathscr{Y}_{11}^r$, $\mathscr{Y}_{12}^r$, $\mathscr{Y}_{22}^r$ and $\mathscr{Z}^r$ evaluated at the time-backward differential equations (4.22)–(4.25) requires that

$$
\begin{aligned}
\frac{\mathrm{d}}{\mathrm{d}\varepsilon} \mathscr{E}_{11}^1(\varepsilon) &= A^T(\varepsilon) \mathscr{H}_{11}^1(\varepsilon) + \mathscr{H}_{11}^1(\varepsilon) A(\varepsilon) + C^T(\varepsilon) Q(\varepsilon) C(\varepsilon) \\
&\quad - \sum_{r=1}^{k} \hat{\mu}_r \mathscr{H}_{11}^r(\varepsilon) B(\varepsilon) R^{-1}(\varepsilon) B^T(\varepsilon) \mathscr{H}_{11}^1(\varepsilon) - \mathscr{H}_{11}^1(\varepsilon) B(\varepsilon) R^{-1}(\varepsilon) B^T(\varepsilon) \\
&\quad \times \sum_{r=1}^{k} \hat{\mu}_r \mathscr{H}_{11}^r(\varepsilon) + \sum_{r=1}^{k} \hat{\mu}_r \mathscr{H}_{11}^r(\varepsilon) B(\varepsilon) R^{-1}(\varepsilon) B^T(\varepsilon) \sum_{s=1}^{k} \hat{\mu}_s \mathscr{H}_{11}^s(\varepsilon),
\end{aligned} \tag{4.36}
$$

$$
\begin{aligned}
\frac{\mathrm{d}}{\mathrm{d}\varepsilon} \mathscr{E}_{11}^r(\varepsilon) &= A^T(\varepsilon) \mathscr{H}_{11}^r(\varepsilon) + \mathscr{H}_{11}^r(\varepsilon) A(\varepsilon) \\
&\quad - \sum_{r=1}^{k} \hat{\mu}_r \mathscr{H}_{11}^r(\varepsilon) B(\varepsilon) R^{-1}(\varepsilon) B^T(\varepsilon) \mathscr{H}_{11}^r(\varepsilon) \\
&\quad - \mathscr{H}_{11}^r(\varepsilon) B(\varepsilon) R^{-1}(\varepsilon) B^T(\varepsilon) \sum_{s=1}^{k} \hat{\mu}_s \mathscr{H}_{11}^s(\varepsilon) \\
&\quad + \sum_{s=1}^{r-1} \frac{2r!}{s!(r-s)!} \mathscr{H}_{11}^s(\varepsilon) G(\varepsilon) W G^T(\varepsilon) \mathscr{H}_{11}^{r-s}(\varepsilon),
\end{aligned} \tag{4.37}
$$

$$
\begin{aligned}
\frac{\mathrm{d}}{\mathrm{d}\varepsilon} \mathscr{E}_{12}^1(\varepsilon) &= A^T(\varepsilon) \mathscr{H}_{12}^1(\varepsilon) + \mathscr{H}_{12}^1(\varepsilon) A_{\mathrm{d}}(\varepsilon) - C^T(\varepsilon) Q(\varepsilon) C_{\mathrm{d}}(\varepsilon) \\
&\quad - \sum_{r=1}^{k} \hat{\mu}_r \mathscr{H}_{11}^r(\varepsilon) B(\varepsilon) R^{-1}(\varepsilon) B^T(\varepsilon) \mathscr{H}_{12}^1(\varepsilon) - \mathscr{H}_{11}^1(\varepsilon) B(\varepsilon) R^{-1}(\varepsilon) B^T(\varepsilon) \\
&\quad \times \sum_{r=1}^{k} \hat{\mu}_r \mathscr{H}_{12}^r(\varepsilon) + \sum_{r=1}^{k} \hat{\mu}_r \mathscr{H}_{11}^r(\varepsilon) B(\varepsilon) R^{-1}(\varepsilon) B^T(\varepsilon) \sum_{s=1}^{k} \hat{\mu}_s \mathscr{H}_{12}^s(\varepsilon),
\end{aligned} \tag{4.38}
$$

$$
\begin{aligned}
\frac{\mathrm{d}}{\mathrm{d}\varepsilon} \mathscr{E}_{12}^r(\varepsilon) &= A^T(\varepsilon) \mathscr{H}_{12}^r(\varepsilon) + \mathscr{H}_{12}^r(\varepsilon) A_{\mathrm{d}}(\varepsilon) \\
&\quad - \sum_{r=1}^{k} \hat{\mu}_r \mathscr{H}_{11}^r(\varepsilon) B(\varepsilon) R^{-1}(\varepsilon) B^T(\varepsilon) \mathscr{H}_{12}^r(\varepsilon)
\end{aligned}
$$

$$- \mathscr{H}_{11}^{r}(\varepsilon)B(\varepsilon)R^{-1}(\varepsilon)B^{T}(\varepsilon)\sum_{r=1}^{k}\hat{\mu}_{r}\mathscr{H}_{12}^{r}(\varepsilon)$$

$$+ \sum_{s=1}^{r-1}\frac{2r!}{s!(r-s)!}\mathscr{H}_{11}^{s}(\varepsilon)G(\varepsilon)WG^{T}(\varepsilon)\mathscr{H}_{12}^{r-s}(\varepsilon), \quad (4.39)$$

$$\frac{\mathrm{d}}{\mathrm{d}\varepsilon}\mathscr{E}_{22}^{1}(\varepsilon) = A_{\mathrm{d}}^{T}(\varepsilon)\mathscr{H}_{22}^{1}(\varepsilon) + \mathscr{H}_{22}^{1}(\varepsilon)A_{\mathrm{d}}(\varepsilon) + C_{\mathrm{d}}^{T}(\varepsilon)Q(\varepsilon)C_{\mathrm{d}}(\varepsilon)$$

$$- \sum_{r=1}^{k}\hat{\mu}_{r}\mathscr{H}_{12}^{r}(\varepsilon)B(\varepsilon)R^{-1}(\varepsilon)B^{T}(\varepsilon)\mathscr{H}_{12}^{1}(\varepsilon)$$

$$- (\mathscr{H}_{12}^{1})^{T}(\varepsilon)B(\varepsilon)R^{-1}(\varepsilon)B^{T}(\varepsilon)\sum_{r=1}^{k}\hat{\mu}_{r}\mathscr{H}_{12}^{r}(\varepsilon)$$

$$+ \sum_{r=1}^{k}\hat{\mu}_{r}\mathscr{H}_{12}^{r}(\varepsilon)B(\varepsilon)R^{-1}(\varepsilon)B^{T}(\varepsilon)\sum_{s=1}^{k}\hat{\mu}_{s}\mathscr{H}_{12}^{s}(\varepsilon), \quad (4.40)$$

$$\frac{\mathrm{d}}{\mathrm{d}\varepsilon}\mathscr{E}_{22}^{r}(\varepsilon) = A_{\mathrm{d}}^{T}(\varepsilon)\mathscr{H}_{22}^{r}(\varepsilon) + \mathscr{H}_{22}^{r}(\varepsilon)A_{\mathrm{d}}(\varepsilon)$$

$$- \sum_{r=1}^{k}\hat{\mu}_{r}\mathscr{H}_{12}^{r}(\varepsilon)B(\varepsilon)R^{-1}(\varepsilon)B^{T}(\varepsilon)\mathscr{H}_{12}^{r}(\varepsilon)$$

$$- (\mathscr{H}_{12}^{r})^{T}(\varepsilon)B(\varepsilon)R^{-1}(\varepsilon)B^{T}(\varepsilon)\sum_{r=1}^{k}\hat{\mu}_{r}\mathscr{H}_{12}^{r}(\varepsilon)$$

$$+ \sum_{s=1}^{r-1}\frac{2r!}{s!(r-s)!}(\mathscr{H}_{12}^{s})^{T}(\varepsilon)G(\varepsilon)WG^{T}(\varepsilon)\mathscr{H}_{12}^{r-s}(\varepsilon), \quad (4.41)$$

$$\frac{\mathrm{d}}{\mathrm{d}\varepsilon}\mathscr{T}^{r}(\varepsilon) = \mathrm{Tr}\left\{\mathscr{H}_{11}^{r}(\varepsilon)G(\varepsilon)WG^{T}(\varepsilon)\right\}. \quad (4.42)$$

The boundary condition of $\mathscr{W}(\varepsilon, \mathscr{Y}_{11}, \mathscr{Y}_{12}, \mathscr{Y}_{22}, \mathscr{Z})$ implies that the initial-value conditions $\mathscr{E}_{11}^{r}(t_0) = 0$, $\mathscr{E}_{12}^{r}(t_0) = 0$, $\mathscr{E}_{22}^{r}(t_0) = 0$ and $\mathscr{T}^{r}(t_0) = 0$ for the forward-in-time differential equations (4.36)–(4.42) and yields a value function

$$\mathscr{W}(\varepsilon, \mathscr{Y}_{11}, \mathscr{Y}_{12}, \mathscr{Y}_{22}, \mathscr{Z}) = \mathscr{V}(\varepsilon, \mathscr{Y}_{11}, \mathscr{Y}_{12}, \mathscr{Y}_{22}, \mathscr{Z})$$

$$= x_0^{T}\sum_{r=1}^{k}\mu_{r}\mathscr{H}_{11}^{r}(t_0)x_0 + 2x_0^{T}\sum_{r=1}^{k}\mu_{r}\mathscr{H}_{12}^{r}(t_0)x_{d0}$$

$$+ x_{d0}^{T}\sum_{r=1}^{k}\mu_{r}\mathscr{H}_{22}^{r}(t_0)x_{d0} + \sum_{r=1}^{k}\mu_{r}\mathscr{D}^{r}(t_0)$$

for which the sufficient condition (4.28) of the verification theorem is satisfied so that the extremizing gains (4.33) and (4.34) become optimal

$$K_x^*(\varepsilon) = -R^{-1}(\varepsilon)B^T(\varepsilon)\sum_{r=1}^{k}\hat{\mu}_r\mathscr{H}_{11}^{*r}(\varepsilon),$$

$$K_{x_d}^*(\varepsilon) = -R^{-1}(\varepsilon)B^T(\varepsilon)\sum_{r=1}^{k}\hat{\mu}_r\mathscr{H}_{12}^{*r}(\varepsilon).$$

Theorem 4.4.1 (Risk-Averse Control Solution for Servo Problems). *Under the assumptions of (A,B) uniformly stabilizable and (C_d, A_d) uniformly detectable, the servo dynamics governed by Eqs. (4.1)–(4.2) follow the desired trajectory of the system (4.3)–(4.4) in accordance of the chi-squared measure of performance (4.5). Suppose $k \in \mathbb{Z}^+$ and the sequence $\mu = \{\mu_r \geq 0\}_{r=1}^k$ with $\mu_1 > 0$ are fixed. Then, the statistical optimal control solution for the servo problem over $[t_0, t_f]$ is a two-degrees-of-freedom controller with time-varying gains*

$$u^*(t) = K_x^*(t)x^*(t) + K_{x_d}^*(t)x_d(t)\,, \quad t = t_0 + t_f - \tau, \tag{4.43}$$

$$K_x^*(\tau) = -R^{-1}(\tau)B^T(\tau)\sum_{r=1}^{k}\hat{\mu}_r\mathscr{H}_{11}^{r*}(\tau), \tag{4.44}$$

$$K_{x_d}^*(\tau) = -R^{-1}(\tau)B^T(\tau)\sum_{r=1}^{k}\hat{\mu}_r\mathscr{H}_{12}^{r*}(\tau), \tag{4.45}$$

whereby $\hat{\mu}_r = \mu_r/\mu_1$ represent different levels of influence as they deem important to the performance distribution.

There is the addition of a feedforward part, which is the state x_d of the reference model (4.3)–(4.4). The feedback part of the optimal servo is dependent on A_d, C_d, and $x_d(t_0)$. Finally, $\{\mathscr{H}_{11}^{r*}(\tau)\}_{r=1}^k$, and $\{\mathscr{H}_{12}^{r*}(\tau)\}_{r=1}^k$ are the optimal solutions of the backward-in-time matrix-valued differential equations

$$\begin{aligned}\frac{d}{d\tau}\mathscr{H}_{11}^{1*}(\tau) = &-[A(\tau)+B(\tau)K_x^*(\tau)]^T\mathscr{H}_{11}^{1*}(\tau) - \mathscr{H}_{11}^{1*}(\tau)[A(\tau)+B(\tau)K_x^*(\tau)]\\ &- K_x^{*T}(\tau)R(\tau)K_x^*(\tau) - C^T(\tau)Q(\tau)C(\tau),\end{aligned} \tag{4.46}$$

$$\begin{aligned}\frac{d}{d\tau}\mathscr{H}_{11}^{r*}(\tau) = &-[A(\tau)+B(\tau)K_x^*(\tau)]^T\mathscr{H}_{11}^{r*}(\tau) - \mathscr{H}_{11}^{r*}(\tau)[A(\tau)+B(\tau)K_x^*(\tau)]\\ &- \sum_{s=1}^{r-1}\frac{2r!}{s!(r-s)!}H_{11}^{s*}(\tau)G(\tau)WG^T(\tau)H_{11}^{(r-s)*}(\tau)\end{aligned} \tag{4.47}$$

and

$$\frac{d}{d\tau}\mathscr{H}_{12}^{1*}(\tau) = -\left[A(\tau)+B(\tau)K_x^*(\tau)\right]^T \mathscr{H}_{12}^{1*}(\tau) - \mathscr{H}_{11}^{1*}(\tau)B(\tau)K_{x_d}^*(\tau) - \mathscr{H}_{12}^{1*}(\tau)A_d(\tau) - K_x^{*T}(\tau)R(\tau)K_{x_d}^*(\tau) + C^T(\tau)Q(\tau)C_d(\tau), \quad (4.48)$$

$$\frac{d}{d\tau}\mathscr{H}_{12}^{r*}(\tau) = -\left[A(\tau)+B(\tau)K_x^*(\tau)\right]^T \mathscr{H}_{12}^{r*}(\tau) - \mathscr{H}_{11}^{r*}(\tau)B(\tau)K_{x_d}^*(\tau) - \mathscr{H}_{12}^{r*}(\tau)A_d(\tau) - \sum_{s=1}^{r-1}\frac{2r!}{s!(r-s)!}H_{11}^{s*}(\tau)G(\tau)WG^T(\tau)H_{12}^{(r-s)*}(\tau) \quad (4.49)$$

whereby the terminal-value conditions $\mathscr{H}_{11}^{1*}(t_f) = C^T(t_f)Q_fC(t_f)$, $\mathscr{H}_{11}^{r*}(t_f) = 0$ *for* $2 \le r \le k$ *as well as* $\mathscr{H}_{12}^{1*}(t_f) = -C^T(t_f)Q_fC_d(t_f)$, $\mathscr{H}_{12}^{r*}(t_f) = 0$ *for* $2 \le r \le k$.

4.5 Chapter Summary

The results here demonstrate a successful combination of the compactness offered by logic from the state-space model description (4.7)–(4.8) and the quantitativity from a-priori probabilistic knowledge of adverse environmental disturbances so that the uncertainty of performance-measure (4.7) can now be represented in a compact and robust way. Moreover, there is a feature of interactive learning in the context of performance uncertainty where the servo controller not only optimizes criteria for evaluating for its performance but also interacts with the external environment. To be specific, the model-reference system (4.8) consists of a statistical controller plus the stochastic elements. It is assumed that the statistical controller has a finite set of performance-measure statistics, while the stochastic elements have a finite set of sample path realizations. The optimal statistical controller composed of cumulant-based feedback (4.44) and feedforward (4.45) gains that operates dynamically on the time-backward histories of the cumulant-supporting equations (4.46)–(4.47) and (4.48)–(4.49) from the final to the current time. Finally, the present framework emphasizes the amount of information in performance-measure statistics (which actually are functions of noise process characteristics) needed to implement a risk-averse learning rule that effectively shapes the closed-loop performance distribution beyond the long-run average performance. The results also stress on the limits of what can be achieved: in such robust control ,designs there exists a kind of statistical controllers (4.43) that trades the property of certainty equivalence principle as would be inherited from the special case of classical linear-quadratic tracking, for the adaptability to deal with uncertain environments.

References

1. Pham, K.D.: New risk-averse control paradigm for stochastic two-time-scale systems and performance robustness. In: Miele, A. (ed.) J. Optim. Theor. Appl. **146**(2), 511–537 (2010)
2. Pham, K.D.: Cost cumulant-based control for a class of linear-quadratic tracking problems. In: Proceedings of the American Control Conference, pp. 335–340 (2007)
3. Pham, K.D., Jin, G., Sain, M.K., Spencer, B.F. Jr., Liberty, S.R.: Generalized LQG techniques for the wind benchmark problem. Special Issue of ASCE J. Eng. Mech. Struct. Contr. Benchmark Prob. **130**(4), 466–470 (2004)
4. Pham, K.D., Sain, M.K., Liberty, S.R.: Cost cumulant control: state-feedback, finite-horizon paradigm with application to seismic protection. In: Miele, A. (ed.) Special issue of J. Optim. Theor. Appl. **115**(3), 685–710 (2002), Kluwer Academic/Plenum Publishers, New York
5. Davison, E.J.: The feedforward control of linear multivariable time-invariant systems. Automatica **9**, 561–573 (1973)
6. Davison, E.J.: The steady-state invertibility and feedforward control of linear time-invariant systems. IEEE Trans. Automat. Contr. **21**, 529–534 (1976)
7. Yuksel, S., Hindi, H., Crawford, L.: Optimal tracking with feedback-feedforward control separation over a network. In: Proceedings of the American Control Conference, pp. 3500–3506 (2006)
8. Fleming, W.H., Rishel, R.W.: Deterministic and Stochastic Optimal Control. Springer, New York (1975)
9. Pham, K.D.: Performance-reliability aided decision making in multiperson quadratic decision games against jamming and estimation confrontations. In: Giannessi, F. (ed.) J. Optim. Theor. Appl. **149**(3), 559–629 (2011)

Chapter 5
Risk-Averse Control Problems in Model-Following Systems

Abstract This chapter considers performance information in a linear-quadratic class of model-following control systems. The innovative idea is the fact that performance information with higher-order performance-measure statistics can improve control decisions for closed-loop system performance reliability but the controller design can also be computationally involved. Many of the results entail measures of the amount, value, and cost of performance information, and the design of model-following control strategy with risk aversion. It becomes clear that the topic of performance information in control is of central importance for future research and development of correct-by-design of high-performance and reliable systems.

5.1 Introduction

Almost all the designers of control systems have included "information process" and "decision process" in their designs as separate items, although many have pointed out that there is no clear-cut separation between these two processes, e.g., the former provides data for the decisions taken by the latter. Most of the work in traditional control theory and stochastic control theory focuses on the design of a decision process in which the information process is either given or chosen arbitrarily. The present research investigation is concerned with the information process which should now be exploited and thus interpreted as "to forecast and provide means examining the future system performance and devising the risk-averse plan of action for performance reliability."

In the recent work [1], the class of decisions implementing risk-averse control strategy pertains to stochastic servo systems on a finite horizon. The novel controller concept was dependent upon the performance information to establish a two-degrees-of-freedom description of desired feedback and feedforward actions for minimal differences between reference and system outputs and trade-offs between risks and benefits for performance reliability. Yet, the results therein were related

K.D. Pham, *Linear-Quadratic Controls in Risk-Averse Decision Making*, SpringerBriefs in Optimization, DOI 10.1007/978-1-4614-5079-5_5,

well with the extensive literature on tracking and feedforward control design problems such as the earlier works of [2–6], just to name a few whose standards have mainly been long-run average performance.

In this chapter, some important results that are new in stochastic control of model-following systems are organized as follows. Section 5.2 characterizes performance uncertainty and risk subject to underlying stochastic disturbances as applied to the problem of structuring and measuring performance variations. Section 5.3 is devoted to the problem statements in model-following systems. It shows a new paradigm of thinking with which the dynamics of cost cumulants can be incorporated into reliable control design with risk-value preferences. Section 5.4 presents a complete development of risk-averse control strategy with the three-degrees-of-freedom structure within which the system not only follows the outputs of the reference model driven by a class of command inputs but also takes into account of performance value and risk trade-offs. Conclusions and future outlooks are also in Sect. 5.5.

5.2 Performance Information in Control with Risk Consequences

Let $(\Omega, \mathscr{F}, \mathbb{F}, \mathbb{P})$ be a complete filtered probability space over $[t_0, t_f]$, on which a p-dimensional standard stationary Wiener process $w(\cdot)$ with the correlation of increments $E\left\{[w(\tau)-w(\xi)][w(\tau)-w(\xi)]^T\right\} = W|\tau-\xi|$ and $W>0$ is defined with $\mathbb{F} = \{\mathscr{F}_t\}_{t \geq t_0 \geq 0}$ being its natural filtration, augmented by all $\mathbb{P}$-null sets in $\mathscr{F}$. Let us first consider a class of stabilizable systems described by controlled stochastic differential equations with the initial system states $x(t_0) = x_0$ known

$$\mathrm{d}x(t) = (A(t)x(t) + B(t)u(t))\,\mathrm{d}t + G(t)\mathrm{d}w(t), \tag{5.1}$$

whereby the continuous-time coefficients $A \in \mathscr{C}([t_0,t_f];\mathbb{R}^{n\times n})$, $B \in \mathscr{C}([t_0,t_f];\mathbb{R}^{n\times m})$, and $G \in \mathscr{C}([t_0,t_f];\mathbb{R}^{n\times p})$. In the above, $x(\cdot)$ is the controlled state process valued in $\mathbb{R}^n$ and $u(\cdot)$ is the control process valued in some set $U \subseteq \mathbb{R}^m$. The output involves some functions $C \in \mathscr{C}([t_0,t_f];\mathbb{R}^{r\times n})$, relating $y(\cdot)$ to the values of the manipulated process $x(\cdot)$

$$y(t) = C(t)x(t)\,. \tag{5.2}$$

In model-following control design, the desired behavior of the output (5.2) is expressed through the use of a reference model driven by a reference input. Typically linear models are used. For instance, a reference model is of the form

$$\mathrm{d}z_1(t) = (A_1(t)z_1(t) + B_1(t)u_r(t))\,\mathrm{d}t\,, \qquad z_1(t_0) = z_{10}, \tag{5.3}$$

$$y_1(t) = C_1(t)z_1(t)\,. \tag{5.4}$$

This reference model is asymptotically stable when $u_r(\cdot) = 0$ and it is controllable with respect to the reference input $u_r(\cdot)$. Furthermore, the reference input $u_r(\cdot)$, valued in $U_r \subseteq \mathbb{R}^{m_1}$, which is also known as the command output of the class of zero input responses of the system

$$dz_2(t) = A_2(t)z_2(t)dt\,, \qquad z_2(t_0) = z_{20}, \tag{5.5}$$

$$u_r(t) = C_2(t)z_2(t), \tag{5.6}$$

whereby some knowledge of continuous-time coefficients $A_1 \in \mathscr{C}([t_0,t_f];\mathbb{R}^{n_1\times n_1})$, $B_1 \in \mathscr{C}([t_0,t_f];\mathbb{R}^{n_1\times m_1})$, $C_1 \in \mathscr{C}([t_0,t_f];\mathbb{R}^{r\times n_1})$ together with $A_2 \in \mathscr{C}([t_0,t_f];\mathbb{R}^{n_2\times n_2})$ and $C_2 \in \mathscr{C}([t_0,t_f];\mathbb{R}^{m_1\times n_2})$ are required.

The purpose of the model-following control design is to use the control process $u(\cdot)$ to force the output process $y(\cdot)$ to track the desired process $y_1(\cdot)$ in spite of the disturbances $w(\cdot)$ and without dependence on the initial states x_0. Associated with the admissible initial condition and control input $(t_0,x_0;u) \in [t_0,t_f] \times \mathbb{R}^n \times U$ is a finite-horizon integral-quadratic-form (IQF) cost $J: [t_0,t_f] \times \mathbb{R}^n \times U \mapsto \mathbb{R}^+$ used for matching of the outputs

$$\begin{aligned} J(t_0,x_0;u) &= (y(t_f) - y_1(t_f))^T Q_f (y(t_f) - y_1(t_f)) \\ &\quad + \int_{t_0}^{t_f} [(y(\tau) - y_1(\tau))^T Q(\tau)(y(\tau) - y_1(\tau)) + u^T(\tau)R(\tau)u(\tau)]d\tau, \end{aligned} \tag{5.7}$$

where the terminal matching penalty $Q_f \in \mathbb{R}^r$, matching penalty $Q \in \mathscr{C}([t_0,t_f];\mathbb{R}^r)$, and the control effort weighting $R \in \mathscr{C}([t_0,t_f];\mathbb{R}^m)$ are symmetric and positive semidefinite matrices with $R(t)$ invertible.

Notice that the model-following problem (5.1)–(5.7) is linear quadratic in nature, the search for closed-loop feedback laws is reasonably restricted within the strategy space which permits a linear feedback synthesis, e.g., $\gamma: [t_0,t_f] \times \mathbb{R}^n \mapsto U$

$$u(t) \triangleq \gamma(t,x(t)) = K_x(t)x(t) + K_{z_1}(t)z_1(t) + v(t), \tag{5.8}$$

where $v(\cdot) \in V \subseteq \mathbb{R}^m$ is an affine control input and while the elements of $K_x \in \mathscr{C}([t_0,t_f];\mathbb{R}^{m\times n})$ and $K_{z_1} \in \mathscr{C}([t_0,t_f];\mathbb{R}^{m\times n_1})$ are admissible feedback, and feedforward gains defined in appropriate senses.

Henceforth, for the admissible initial condition $(t_0,x_0,z_{10},z_{20}) \in [t_0,t_f] \times \mathbb{R}^n \times \mathbb{R}^{n_1} \times \mathbb{R}^{n_2}$ and the feasible control (5.8), the aggregation of Eqs. (5.1) and (5.3) is described by the controlled stochastic differential equation

$$dm(t) = (A_a(t)m(t) + B_a(t)v(t) + \Gamma_a(t)u_r(t))\,dt + G_a(t)dw(t),\, m(t_0) = m_0 \tag{5.9}$$

with the performance measure (5.7) rewritten as follows

$$\begin{aligned} J = m^T(t_f)Q_{fa}m(t_f) &+ \int_{t_0}^{t_f} \big[m^T(\tau)Q_a(\tau)m(\tau) \\ &+ 2m^T(\tau)N_a(\tau)v(\tau) + v^T(\tau)R(\tau)v(\tau)\big]\, d\tau, \end{aligned} \tag{5.10}$$

whereby the aggregate state variables and continuous-time coefficients are given by

$$m = \begin{bmatrix} x \\ z_1 \end{bmatrix} \quad m_0 = \begin{bmatrix} x_0 \\ z_{10} \end{bmatrix}, \quad A_a = \begin{bmatrix} A+BK_x & BK_{z_1} \\ 0 & A_1 \end{bmatrix}, \quad B_a = \begin{bmatrix} B \\ 0 \end{bmatrix}$$

$$\Gamma_a = \begin{bmatrix} 0 \\ B_1 \end{bmatrix}, \quad G_a = \begin{bmatrix} G \\ 0 \end{bmatrix}, \quad Q_{fa} = \begin{bmatrix} C^T(t_f)Q_fC(t_f) & -C^T(t_f)Q_fC_1(t_f) \\ -C_1^T(t_f)Q_fC(t_f) & C_1^T(t_f)Q_fC_1(t_f) \end{bmatrix}$$

$$N_a = \begin{bmatrix} K_x^T R \\ K_{z_1}^T R \end{bmatrix}, \quad Q_a = \begin{bmatrix} C^TQC + K_x^TRK_x & -C^TQC_1 + K_x^TRK_{z_1} \\ -C_1^TQC + K_{z_1}^TRK_x & C_1^TQC_1 + K_{z_1}^TRK_{z_1} \end{bmatrix}.$$

So far there are two types of information, i.e., process information (5.9) and goal information (5.10), which have been given in advance to the controller (5.8) Since there is an external disturbance $w(\cdot)$ affecting the closed-loop performance, the controller now needs additional information about performance variations. This is *coupling information* and thus also known as *performance information*. The questions of how to characterize and influence performance information are then answered by adaptive cost cumulants (also known as semi-invariants) associated with the performance measure (5.10) in detail below. Precisely stated, these cost cumulants can be generated by parameterizing the initial condition (t_0, m_0) to any arbitrary pair (τ, m_τ). Then, for the given affine signal v, admissible feedback K_x, and feedforward K_{z_1}, the performance measure (5.10) is seen as the "cost-to-go", $J(\tau, m_\tau)$. The moment-generating function of the performance measure (5.10) is defined by

$$\varphi(\tau, m_\tau; \theta) \triangleq E\{\exp(\theta J(\tau, m_\tau))\}, \tag{5.11}$$

whereby the scalar $\theta \in \mathbb{R}^+$ is a small parameter. Thus, the cumulant-generating function immediately follows

$$\psi(\tau, m_\tau; \theta) \triangleq \ln\{\varphi(\tau, m_\tau; \theta)\}, \tag{5.12}$$

whereby $\ln\{\cdot\}$ is a natural logarithmic transformation.

Theorem 5.2.1 (Cumulant-Generating Function). *Let the system be governed by Eqs. (5.9)–(5.10). Let the sequel definitions be as follows:*

$$\varphi(\tau, m_\tau; \theta) \triangleq \rho_a(\tau, \theta) \exp\{m_\tau^T \Upsilon_a(\tau, \theta) m_\tau + 2m_\tau^T \eta_a(\tau, \theta)\}$$

$$\upsilon_a(\tau, \theta) \triangleq \ln\{\rho_a(\tau, \theta)\}, \quad \theta \in \mathbb{R}^+, \quad \tau \in [t_0, t_f].$$

Then, the cumulant-generating function is computed as

$$\psi(\tau, m_\tau; \theta) = m_\tau^T \Upsilon_a(\tau, \theta) m_\tau + 2m_\tau^T \eta_a(\tau, \theta) + \upsilon_a(\tau, \theta) \tag{5.13}$$

whereby the cumulant-supporting variables $\Upsilon_a(\tau, \theta)$, $\eta_a(\tau, \theta)$ and $\upsilon_a(\tau, \theta)$ solve the backward-in-time matrix-valued differential equation

$$\frac{d}{d\tau}\Upsilon_a(\tau,\theta) = -A_a^T(\tau)\Upsilon_a(\tau,\theta) - \Upsilon_a(\tau,\theta)A_a(\tau) - \theta Q_a(\tau)$$
$$- 2\Upsilon_a(\tau,\theta)G_a(\tau)WG_a^T(\tau)\Upsilon_a(\tau,\theta), \quad \Upsilon_a(t_f,\theta) = \theta Q_{fa} \qquad (5.14)$$

the backward-in-time vector-valued differential equation

$$\frac{d}{d\tau}\eta_a(\tau,\theta) = -A_a^T(\tau)\eta_a(\tau,\theta) - \Upsilon_a(\tau,\theta)B_a(\tau)v(\tau)$$
$$- \Upsilon_a(\tau,\theta)\Gamma_a(\tau)u_r(\tau) - \theta N_a(\tau)v(\tau), \quad \eta_a(t_f,\theta) = 0 \qquad (5.15)$$

and the backward-in-time scalar-valued differential equation

$$\frac{d}{d\tau}\upsilon_a(\tau,\theta) = -\mathrm{Tr}\{\Upsilon_a(\tau,\theta)G_a(\tau)WG_a^T(\tau)\} - \theta v^T(\tau)R(\tau)v(\tau)$$
$$- 2\eta_a^T(\tau,\theta)(B_a(\tau)v(\tau) + \Gamma_a(\tau)u_r(\tau)), \quad \upsilon_a(t_f,\theta) = 0. \qquad (5.16)$$

Proof. The proof is similar to those of the previous chapters and alternate work by the author available at [7]. □

By definition, performance-measure statistics or cost cumulants for short that provide performance information for the decision process taken by the model-following control design can best be generated by the Maclaurin series expansion of the cumulant-generating function (5.13)

$$\psi(\tau,m_\tau;\theta) \triangleq \sum_{r=1}^{\infty} \kappa_r \frac{\theta^r}{r!} = \sum_{r=1}^{\infty} \left.\frac{\partial^r}{\partial\theta^r}\psi(\tau,m_\tau;\theta)\right|_{\theta=0} \frac{\theta^r}{r!},$$

whereby κ_r are denoted as the rth-order performance-measure statistics or cost cumulants associated with the chi-squared random variable (5.10).

In view of the cumulant-generating function (5.13), all the performance-measure statistics that measure amount of information about performance variations in Eq. (5.10) are obtained as follows

$$\kappa_r = m_\tau^T \left.\frac{\partial^r}{\partial\theta^r}\Upsilon_a(\tau,\theta)\right|_{\theta=0} m_\tau + 2m_\tau^T \left.\frac{\partial^r}{\partial\theta^r}\eta_a(\tau,\theta)\right|_{\theta=0} + \left.\frac{\partial^r}{\partial\theta^r}\upsilon_a(\tau,\theta)\right|_{\theta=0}.$$

For notational convenience, it is necessary to introduce

$$H_a(\tau,r) \triangleq \left.\frac{\partial^r}{\partial\theta^r}\Upsilon_a(\tau,\theta)\right|_{\theta=0}, \qquad D_a(\tau,r) \triangleq \left.\frac{\partial^r}{\partial\theta^r}\eta_a(\tau,\theta)\right|_{\theta=0}$$
$$d_a(\tau,r) \triangleq \left.\frac{\partial^r}{\partial\theta^r}\upsilon_a(\tau,\theta)\right|_{\theta=0}, \qquad r \in \mathbb{Z}^+.$$

Then, the next result whose proof has already been outlined as above, provides measures of the amount, value and performance information structures.

Theorem 5.2.2 (Performance-Measure Statistics). *In spirit of the model-following system (5.9)–(5.10), the kth-order performance-measure statistic associated with the random variable (5.10) is given by*

$$\kappa_k = x_{a0}^T H_a(t_0,k) x_{a0} + 2x_{a0}^T D_a(t_0,k) + d_a(t_0,k), \quad k \in \mathbb{Z}^+ \tag{5.17}$$

whereby the solutions $\{H_a(\tau,r)\}_{r=1}^k$, $\{D_a(\tau,r)\}_{r=1}^k$ *and* $\{d_a(\tau,r)\}_{r=1}^k$ *evaluated at* $\tau = t_0$ *satisfy the backward-in-time differential equations (with the dependence of* $H_a(\tau,r)$, $D_a(\tau,r)$, *and* $d_a(\tau,r)$ *upon* K_x, K_{z_1}, *and* v *suppressed)*

$$\frac{d}{d\tau} H_a(\tau,1) = -A_a^T(\tau) H_a(\tau,1) - H_a(\tau,1) A_a(\tau) - Q_a(\tau), \tag{5.18}$$

$$\frac{d}{d\tau} H_a(\tau,r) = -A_a^T(\tau) H_a(\tau,r) - H_a(\tau,r) A_a(\tau) - \sum_{s=1}^{r-1} \frac{2r!}{s!(r-s)!} H_a(\tau,s) G_a(\tau) W G_a^T(\tau) H_a(\tau,r-s), \quad 2 \le r \le k, \tag{5.19}$$

$$\frac{d}{d\tau} D_a(\tau,1) = -A_a^T(\tau) D_a(\tau,1) - H_a(\tau,1) B_a(\tau) v(\tau) - H_a(\tau,1) \Gamma_a(\tau) u_r(\tau) - N_a(\tau) v(\tau), \tag{5.20}$$

$$\frac{d}{d\tau} D_a(\tau,r) = -A_a^T(\tau) D_a(\tau,r) - H_a(\tau,r) B_a(\tau) v(\tau) - H_a(\tau,r) \Gamma_a(\tau) u_r(\tau), \quad 2 \le r \le k, \tag{5.21}$$

$$\frac{d}{d\alpha} d_a(\tau,1) = -\mathrm{Tr}\left\{H_a(\tau,1) G_a(\tau) W G_a^T(\tau)\right\} - 2D_a^T(\tau,1)(B_a(\tau) v(\tau) + \Gamma_a(\tau) u_r(\tau)) - v^T(\tau) R(\tau) v(\tau), \tag{5.22}$$

$$\frac{d}{d\alpha} d_a(\tau,r) = -\mathrm{Tr}\left\{H_a(\tau,r) G_a(\tau) W G_a^T(\tau)\right\} - 2D_a^T(\tau,r)(B_a(\tau) v(\tau) + \Gamma_a(\tau) u_r(\tau)), \quad 2 \le r \le k, \tag{5.23}$$

where the terminal-value conditions $H_a(t_f,1) = Q_{fa}$, $H_a(t_f,r) = 0$ *for* $2 \le r \le k$, $D_a(t_f,r) = 0$ *for* $1 \le r \le k$ *and* $d_a(t_f,r) = 0$ *for* $1 \le r \le k$.

In the design of a decision process in which the information process about performance variations is embedded with K_x, K_{z_1}, and v, it is convenient to rewrite the results (5.17)–(5.23) in accordance of the matrix and vector partitions

$$H_a(\tau,r) = \begin{bmatrix} \mathscr{H}_{11}^r(\tau) & \mathscr{H}_{12}^r(\tau) \\ (\mathscr{H}_{12}^r)^T(\tau) & \mathscr{H}_{22}^r(\tau) \end{bmatrix} \text{ and } D_a(\tau,r) = \begin{bmatrix} \mathscr{D}_{11}^r(\tau) \\ \mathscr{D}_{21}^r(\tau) \end{bmatrix}.$$

Corollary 5.2.1 (Alternate Representation of Performance-Measure Statistics). *Let the model-following system be described by Eqs. (5.9)–(5.10). Then, the kth-order performance-measure statistics of the chi-squared random variable (5.10) can also be rewritten as follows:*

$$\begin{aligned}\kappa_k &= x_0^T \mathscr{H}_{11}^k(t_0)x_0 + 2x_0^T \mathscr{H}_{12}^k(t_0)z_{10} + z_{10}^T \mathscr{H}_{22}^k(t_0)z_{10} \\ &\quad + \mathscr{D}^k(t_0) + 2x_0^T \mathscr{D}_{11}^k(t_0) + 2z_{10}^T \mathscr{D}_{21}^k(t_0)\,, \quad k \in \mathbb{Z}^+ \end{aligned} \tag{5.24}$$

where the cumulant-supporting variables $\{\mathscr{H}_{11}^r(\tau)\}_{r=1}^k$, $\{\mathscr{H}_{12}^r(\tau)\}_{r=1}^k$, $\{\mathscr{H}_{22}^r(\tau)\}_{r=1}^k$, $\{\mathscr{D}_{11}^r(\tau)\}_{r=1}^k$, $\{\mathscr{D}_{21}^r(\tau)\}_{r=1}^k$, *and* $\{\mathscr{D}^r(\tau)\}_{r=1}^k$ *satisfy the time-backward differential equations*

$$\begin{aligned}\frac{d}{d\tau}\mathscr{H}_{11}^1(\tau) &\triangleq \mathscr{F}_{11}^1(\tau,\mathscr{H}_{11},K_x) \\ &= -\left[A(\tau)+B(\tau)K_x(\tau)\right]^T \mathscr{H}_{11}^1(\tau) - C^T(\tau)Q(\tau)C(\tau) \\ &\quad - \mathscr{H}_{11}^1(\tau)\left[A(\tau)+B(\tau)K_x(\tau)\right] - K_x^T(\tau)R(\tau)K_x(\tau), \end{aligned} \tag{5.25}$$

$$\begin{aligned}\frac{d}{d\tau}\mathscr{H}_{11}^r(\tau) &\triangleq \mathscr{F}_{11}^r(\tau,\mathscr{H}_{11},K_x) \\ &= -\left[A(\tau)+B(\tau)K_x(\tau)\right]^T \mathscr{H}_{11}^r(\tau) - \mathscr{H}_{11}^r(\tau)\left[A(\tau)+B(\tau)K_x(\tau)\right] \\ &\quad - \sum_{s-1}^{r-1} \frac{2r!}{s!(r-s)!}\mathscr{H}_{11}^s(\tau)G(\tau)WG^T(\tau)\mathscr{H}_{11}^{r-s}(\tau), \end{aligned} \tag{5.26}$$

$$\begin{aligned}\frac{d}{d\tau}\mathscr{H}_{12}^1(\tau) &\triangleq \mathscr{F}_{12}^1(\tau,\mathscr{H}_{12},\mathscr{H}_{11},K_x,K_{z_1}) \\ &= -\left[A(\tau)+B(\tau)K_x(\tau)\right]^T \mathscr{H}_{12}^1(\tau) - K_x^T(\tau)R(\tau)K_{z_1}(\tau) \\ &\quad - \mathscr{H}_{11}^1(\tau)B(\tau)K_{z_1}(\tau) - \mathscr{H}_{12}^1(\tau)A_1(\tau) + C^T(\tau)Q(\tau)C_1(\tau), \end{aligned} \tag{5.27}$$

$$\begin{aligned}\frac{d}{d\tau}\mathscr{H}_{12}^r(\tau) &\triangleq \mathscr{F}_{12}^r(\tau,\mathscr{H}_{12},\mathscr{H}_{11},K_x,K_{z_1}) \\ &= -\left[A(\tau)+B(\tau)K_x(\tau)\right]^T \mathscr{H}_{12}^r(\tau) - \mathscr{H}_{11}^r(\tau)B(\tau)K_{z_1}(\tau) - \mathscr{H}_{12}^r(\tau)A_1(\tau) \\ &\quad - \sum_{s=1}^{r-1} \frac{2r!}{s!(r-s)!}\mathscr{H}_{11}^s(\tau)G(\tau)WG^T(\tau)\mathscr{H}_{12}^{r-s}(\tau), \end{aligned} \tag{5.28}$$

$$\begin{aligned}\frac{d}{d\tau}\mathscr{H}_{22}^1(\tau) &\triangleq \mathscr{F}_{22}^1(\tau,\mathscr{H}_{22},\mathscr{H}_{12},K_{z_1}) \\ &= -A_1^T(\tau)\mathscr{H}_{22}^1(\tau) - \mathscr{H}_{22}^1(\tau)A_1(\tau) - K_{z_1}^T(\tau)B^T(\tau)\mathscr{H}_{12}^1(\tau) \\ &\quad - (\mathscr{H}_{12}^1)^T(\tau)B(\tau)K_{z_1}(\tau) - C_1^T(\tau)Q(\tau)C_1(\tau) - K_{z_1}^T(\tau)R(\tau)K_{z_1}(\tau), \end{aligned} \tag{5.29}$$

$$
\begin{aligned}
\frac{d}{d\tau}\mathcal{H}_{22}^r(\tau) &\triangleq \mathcal{F}_{22}^r(\tau,\mathcal{H}_{22},\mathcal{H}_{12},K_{z_1}) \\
&= -A_1^T(\tau)\mathcal{H}_{22}^r(\tau) - \mathcal{H}_{22}^r(\tau)A_1(\tau) - K_{z_1}^T(\tau)B^T(\tau)\mathcal{H}_{12}^r(\tau) \\
&\quad - (\mathcal{H}_{12}^r)^T(\tau)B(\tau)K_{z_1}(\tau) \\
&\quad - \sum_{s=1}^{r-1}\frac{2r!}{s!(r-s)!}(\mathcal{H}_{12}^s)^T(\tau)G(\tau)WG^T(\tau)\mathcal{H}_{12}^{r-s}(\tau), \qquad (5.30)
\end{aligned}
$$

$$
\begin{aligned}
\frac{d}{d\tau}\mathcal{D}_{11}^1(\tau) &\triangleq \mathcal{G}_{11}^1(\tau,\mathcal{D}_{11},\mathcal{H}_{11},\mathcal{H}_{12},K_x,v) \\
&= -\left[A(\tau)+B(\tau)K_x(\tau)\right]^T\mathcal{D}_{11}^1(\tau) - \mathcal{H}_{11}^1(\tau)B(\tau)v(\tau) \\
&\quad - \mathcal{H}_{12}^1(\tau)B_1(\tau)u_r(\tau) - K_x^T(\tau)R(\tau)v(\tau), \qquad (5.31)
\end{aligned}
$$

$$
\begin{aligned}
\frac{d}{d\tau}\mathcal{D}_{11}^r(\tau) &\triangleq \mathcal{G}_{11}^r(\tau,\mathcal{D}_{11},\mathcal{H}_{11},\mathcal{H}_{12},K_x,v) \\
&= -\left[A(\tau)+B(\tau)K_x(\tau)\right]^T\mathcal{D}_{11}^r(\tau) - \mathcal{H}_{11}^r(\tau)B(\tau)v(\tau) \\
&\quad - \mathcal{H}_{12}^r(\tau)B_1(\tau)u_r(\tau), \quad 2\le r\le k, \qquad (5.32)
\end{aligned}
$$

$$
\begin{aligned}
\frac{d}{d\tau}\mathcal{D}_{21}^1(\tau) &\triangleq \mathcal{G}_{21}^1(\tau,\mathcal{D}_{21},\mathcal{D}_{11},\mathcal{H}_{12},\mathcal{H}_{22},K_{z_1},v) \\
&= -K_{z_1}^T(\tau)B^T(\tau)\mathcal{D}_{11}^1(\tau) - (\mathcal{H}_{12}^1)^T(\tau)B(\tau)v(\tau) \\
&\quad - A_1^T(\tau)\mathcal{D}_{21}^1(\tau) - \mathcal{H}_{22}^1(\tau)B_1(\tau)u_r(\tau) - K_{z_1}^T(\tau)R(\tau)v(\tau), \qquad (5.33)
\end{aligned}
$$

$$
\begin{aligned}
\frac{d}{d\tau}\mathcal{D}_{21}^r(\tau) &\triangleq \mathcal{G}_{21}^1(\tau,\mathcal{D}_{21},\mathcal{D}_{11},\mathcal{H}_{12},\mathcal{H}_{22},K_{z_1},v) \\
&= -K_{z_1}^T(\tau)B^T(\tau)\mathcal{D}_{11}^r(\tau) - (\mathcal{H}_{12}^r)^T(\tau)B(\tau)v(\tau) \\
&\quad - A_1^T(\tau)\mathcal{D}_{21}^r(\tau) - \mathcal{H}_{22}^r(\tau)B_1(\tau)u_r(\tau), \quad 2\le r\le k, \qquad (5.34)
\end{aligned}
$$

$$
\begin{aligned}
\frac{d}{d\tau}\mathcal{D}^1(\tau) &\triangleq \mathcal{G}^1(\tau,\mathcal{H}_{11},\mathcal{D}_{11},\mathcal{D}_{21},v) \\
&= -\mathrm{Tr}\left\{\mathcal{H}_{11}^1(\tau)G(\tau)WG^T(\tau)\right\} - 2(\mathcal{D}_{11}^1)^T(\tau)B(\tau)v(\tau) \\
&\quad - 2(\mathcal{D}_{21}^1)^T(\tau)B_1(\tau)u_r(\tau) - v^T(\tau)R(\tau)v(\tau), \qquad (5.35)
\end{aligned}
$$

$$
\begin{aligned}
\frac{d}{d\tau}\mathcal{D}^r(\tau) &\triangleq \mathcal{G}^r(\tau,\mathcal{H}_{11},\mathcal{D}_{11},\mathcal{D}_{21},v) \\
&= -\mathrm{Tr}\left\{\mathcal{H}_{11}^r(\tau)G(\tau)WG^T(\tau)\right\} - 2(\mathcal{D}_{11}^r)^T(\tau)B(\tau)v(\tau) \\
&\quad - 2(\mathcal{D}_{21}^r)^T(\tau)B_1(\tau)u_r(\tau), \quad 2\le r\le k, \qquad (5.36)
\end{aligned}
$$

whereby the terminal-value conditions are $\mathcal{H}_{11}^1(t_f) = C^T(t_f)Q_fC(t_f)$, $\mathcal{H}_{11}^r(t_f) = 0$ *for* $2 \le r \le k$; $\mathcal{H}_{12}^1(t_f) = -C^T(t_f)Q_fC_1(t_f)$, $\mathcal{H}_{12}^r(t_f) = 0$ *for* $2 \le r \le k$; $\mathcal{H}_{22}^1(t_f) = C_1^T(t_f)Q_fC_1(t_f)$, $\mathcal{H}_{22}^r(t_f) = 0$ *for* $2 \le r \le k$; $\mathcal{D}_{11}^r(t_f) = 0$ *for* $1 \le r \le k$; $\mathcal{D}_{21}^r(t_f) = 0$ *for* $1 \le r \le k$; $\mathcal{D}^r(t_f) = 0$ *for* $1 \le r \le k$.

And the components of the k-tuple variables $\mathcal{H}_{11}$, $\mathcal{H}_{12}$, $\mathcal{H}_{22}$, $\mathcal{D}_{11}$, $\mathcal{D}_{21}$, *and* $\mathcal{D}$ *are defined by*

$$\mathcal{H}_{11} \triangleq (\mathcal{H}_{11}^1,\ldots,\mathcal{H}_{11}^k), \quad \mathcal{H}_{12} \triangleq (\mathcal{H}_{12}^1,\ldots,\mathcal{H}_{12}^k), \quad \mathcal{H}_{22} \triangleq (\mathcal{H}_{22}^1,\ldots,\mathcal{H}_{22}^k)$$

$$\mathcal{D}_{11} \triangleq (\mathcal{D}_{11}^1,\ldots,\mathcal{D}_{11}^k), \quad \mathcal{D}_{21} \triangleq (\mathcal{D}_{21}^1,\ldots,\mathcal{D}_{21}^k), \quad \mathcal{D} \triangleq (\mathcal{D}^1,\ldots,\mathcal{D}^k).$$

To anticipate for a well-posed optimization problem that follows, some sufficient conditions for the existence of solutions to the cumulant-generating equations (5.25)–(5.36) are now presented as follows.

Corollary 5.2.2 (Existence of Performance-Measure Statistics). *In addition to asymptotically stable reference models, it is assumed that* (A,B) *and* (A,C) *are uniformly stabilizable and detectable, respectively. Then, for any given* $k \in \mathbb{N}$, *the backward-in-time differential equations (5.25)–(5.36) admit unique and bounded solutions, e.g.,* $\{\mathcal{H}_{11}^r(\cdot)\}_{r=1}^k$, $\{\mathcal{H}_{12}^r(\cdot)\}_{r=1}^k$, $\{\mathcal{H}_{22}^r(\cdot)\}_{r=1}^k$, $\{\mathcal{D}_{11}^r(\cdot)\}_{r=1}^k$, $\{\mathcal{D}_{21}^r(\cdot)\}_{r=1}^k$, *and* $\{\mathcal{D}^r(\cdot)\}_{r=1}^k$ *on* $[t_0,t_f]$.

Proof. The rigorous proof is deferred to the work by the author available at [7]. □

5.3 Formulation of the Control Problem

To formulate in precise terms for optimization problem in the model-following control design, it is important to note that all cumulants (5.24) are the functions of backward-in-time evolutions governed by Eqs. (5.25)–(5.36) and initial valued states $x(t_0)$ and $z_1(t_0)$. Henceforth, these backward-in-time trajectories (5.25)–(5.36) are therefore considered as the new dynamical equations together with state variables $\{\mathcal{H}_{11}^r(\cdot)\}_{r=1}^k$, $\{\mathcal{H}_{12}^r(\cdot)\}_{r=1}^k$, $\{\mathcal{H}_{22}^r(\cdot)\}_{r=1}^k$, $\{\mathcal{D}_{11}^r(\cdot)\}_{r=1}^k$, $\{\mathcal{D}_{21}^r(\cdot)\}_{r=1}^k$ and $\{\mathcal{D}^r(\cdot)\}_{r=1}^k$, and thus not the traditional system states $x(\cdot)$ and $z_1(\cdot)$.

One can infer from the following mathematical definitions on product mappings

$$\mathcal{F}_{11} \triangleq \mathcal{F}_{11}^1 \times \cdots \times \mathcal{F}_{11}^k, \quad \mathcal{F}_{12} \triangleq \mathcal{F}_{12}^1 \times \cdots \times \mathcal{F}_{12}^k, \quad \mathcal{F}_{22} \triangleq \mathcal{F}_{22}^1 \times \cdots \times \mathcal{F}_{22}^k,$$

$$\mathcal{G}_{11} \triangleq \mathcal{G}_{11}^1 \times \cdots \times \mathcal{G}_{11}^k, \quad \mathcal{G}_{21} \triangleq \mathcal{G}_{21}^1 \times \cdots \times \mathcal{G}_{21}^k, \quad \mathcal{G} \triangleq \mathcal{G}^1 \times \cdots \times \mathcal{G}^k$$

that the dynamical equations (5.25)–(5.36) can be rewritten compactly as

$$\frac{d}{d\tau}\mathcal{H}_{11}(\tau) = \mathcal{F}_{11}(\tau,\mathcal{H}_{11},K_x), \quad \mathcal{H}_{11}(t_f) = \mathcal{H}_{11}^f, \tag{5.37}$$

$$\frac{d}{d\tau}\mathcal{H}_{12}(\tau) = \mathcal{F}_{12}(\tau,\mathcal{H}_{12},\mathcal{H}_{11},K_x,K_{z_1}), \quad \mathcal{H}_{12}(t_f) = \mathcal{H}_{12}^f, \tag{5.38}$$

$$\frac{\mathrm{d}}{\mathrm{d}\tau}\mathscr{H}_{22}(\tau) = \mathscr{F}_{22}(\tau, \mathscr{H}_{22}, \mathscr{H}_{12}, K_{z_1}), \quad \mathscr{H}_{22}(t_f) = \mathscr{H}_{22}^f, \tag{5.39}$$

$$\frac{\mathrm{d}}{\mathrm{d}\tau}\mathscr{D}_{11}(\tau) = \mathscr{G}_{11}(\tau, \mathscr{D}_{11}, \mathscr{H}_{11}, \mathscr{H}_{12}, K_x, v), \quad \mathscr{D}_{11}(t_f) = \mathscr{D}_{11}^f, \tag{5.40}$$

$$\frac{\mathrm{d}}{\mathrm{d}\tau}\mathscr{D}_{21}(\tau) = \mathscr{G}_{21}(\tau, \mathscr{D}_{21}, \mathscr{D}_{11}, \mathscr{H}_{12}, \mathscr{H}_{22}, K_{z_1}, v), \quad \mathscr{D}_{21}(t_f) = \mathscr{D}_{21}^f, \tag{5.41}$$

$$\frac{\mathrm{d}}{\mathrm{d}\tau}\mathscr{D}(\tau) = \mathscr{G}(\tau, \mathscr{H}_{11}, \mathscr{D}_{11}, \mathscr{D}_{21}, v), \quad \mathscr{D}(t_f) = \mathscr{D}^f, \tag{5.42}$$

whereby the terminal-value conditions for the k-tuple state variables $\mathscr{H}_{11}(\cdot)$, $\mathscr{H}_{12}(\cdot)$, $\mathscr{H}_{22}(\cdot)$, $\mathscr{D}_{11}(\cdot)$, $\mathscr{D}_{21}(\cdot)$ and $\mathscr{D}(\cdot)$ are given by

$$\mathscr{H}_{11}^f \triangleq (C^T(t_f)Q_f C(t_f), 0, \ldots, 0), \qquad \mathscr{D}_{11}^f \triangleq (0, \ldots, 0),$$

$$\mathscr{H}_{12}^f \triangleq (-C^T(t_f)Q_f C_1(t_f), 0, \ldots, 0), \qquad \mathscr{D}_{21}^f \triangleq (0, \ldots, 0),$$

$$\mathscr{H}_{22}^f \triangleq (C_1^T(t_f)Q_f C_1(t_f), 0, \ldots, 0), \qquad \mathscr{D}^f \triangleq (0, \ldots, 0).$$

Interesting enough, the product system (5.37)–(5.42) uniquely determines the state variables $\mathscr{H}_{11}(\cdot)$, $\mathscr{H}_{12}(\cdot)$, $\mathscr{H}_{22}(\cdot)$, $\mathscr{D}_{11}(\cdot)$, $\mathscr{D}_{21}(\cdot)$, and $\mathscr{D}(\cdot)$ once the admissible feedback gain $K_x(\cdot)$, feedforward gain $K_{z_1}(\cdot)$, and affine input $v(\cdot)$ are specified. Henceforth, the state variables $\mathscr{H}_{11}(\cdot)$, $\mathscr{H}_{12}(\cdot)$, $\mathscr{H}_{22}(\cdot)$, $\mathscr{D}_{11}(\cdot)$, $\mathscr{D}_{21}(\cdot)$ and $\mathscr{D}(\cdot)$ are considered as $\mathscr{H}_{11}(\cdot, K_x, K_{z_1}, v)$, $\mathscr{H}_{12}(\cdot, K_x, K_{z_1}, v)$, $\mathscr{H}_{22}(\cdot, K_x, K_{z_1}, v)$, $\mathscr{D}_{11}(\cdot, K_x, K_{z_1}, v)$, $\mathscr{D}_{21}(\cdot, K_x, K_{z_1}, v)$ and $\mathscr{D}(\cdot, K_x, K_{z_1}, v)$, respectively.

What comes next is a new model concept of performance benefit and risk which consists of tradeoffs between performance values and risks for reliable control of model-following systems. It therefore underlies a risk aversion notion in performance assessment and feedback control design under uncertainty. In a sense, the generalized performance index proposed herein for the model-following control design provides an effective framework for decision and control processes with benefit and risk awareness.

Definition 5.3.1 (Risk-Value Aware Performance Index). Fix $k \in \mathbb{Z}^+$ and the sequence $\mu = \{\mu_r \geq 0\}_{r=1}^k$ with $\mu_1 > 0$. Then, the performance index of Mayer type with value and risk considerations is given by

$$\begin{aligned}
&\phi_0^m(t_0, \mathscr{H}_{11}(t_0), \mathscr{H}_{12}(t_0), \mathscr{H}_{22}(t_0), \mathscr{D}_{11}(t_0), \mathscr{D}_{21}(t_0), \mathscr{D}(t_0)) \\
&\triangleq \underbrace{\mu_1\kappa_1}_{\text{Value Measure}} + \underbrace{\mu_2\kappa_2 + \cdots + \mu_k\kappa_k}_{\text{Risk Measures}} \\
&= \sum_{r=1}^k \mu_r \Big[x_0^T \mathscr{H}_{11}^r(t_0)x_0 + 2x_0^T \mathscr{H}_{12}^r(t_0)z_{10} + z_{10}^T \mathscr{H}_{22}^r(t_0)z_{10} \\
&\quad + 2x_0^T \mathscr{D}_{11}^r(t_0) + 2z_{10}^T \mathscr{D}_{21}^r(t_0) + \mathscr{D}^r(t_0)\Big],
\end{aligned} \tag{5.43}$$

where the parametric design of freedom μ_r chosen by the control designer represent preferential weights on simultaneous higher-order characteristics, e.g., mean (i.e., the average of performance measure), variance (i.e., the dispersion of values of performance measure around its mean), skewness (i.e., the antisymmetry of the probability density of performance measure), and kurtosis (i.e., the heaviness in the probability density tails of performance measure) , pertaining to closed-loop performance variations and uncertainties; whereas the k-tuple cumulant variables $\mathscr{H}_{11}$, $\mathscr{H}_{12}$, $\mathscr{H}_{22}$, $\mathscr{D}_{11}$, $\mathscr{D}_{21}$, and $\mathscr{D}$ evaluated at $\tau = t_0$ satisfy the new dynamics (5.37)–(5.42).

Then, the method of dynamic programming is employed to obtain necessary and sufficient conditions for the existence of an explicit, closed-form solution of the linear-quadratic class of model-following control systems considered here. Henceforth, in view of the performance index with risk aversion (5.43), any candidate value function associated with (5.43) shall only be a function of supporting variables $\mathscr{H}_{11}$, $\mathscr{H}_{12}$, $\mathscr{H}_{22}$, $\mathscr{D}_{11}$, $\mathscr{D}_{21}$ and $\mathscr{D}$ for higher-order performance-measure statistics but does not depend on the traditional states x.

In conformity with the rigorous formulation of dynamic programming, the development that follows is important. For terminal data $(t_f, \mathscr{H}_{11}^f, \mathscr{H}_{12}^f, \mathscr{H}_{22}^f, \mathscr{D}_{11}^f, \mathscr{D}_{21}^f, \mathscr{D}^f)$ given, the classes of admissible feedback, feedforward gains, and affine inputs are then defined as follows.

Definition 5.3.2 (Admissible Gains and Affine Inputs). Let compact subsets $\overline{P}_x \subset \mathbb{R}^{m\times n}$, $\overline{P}_{z_1} \subset \mathbb{R}^{m\times n_1}$, and $\overline{V} \subset \mathbb{R}^m$ be the allowable sets of gains and affine input values. For the given $k \in \mathbb{Z}^+$ and the sequence $\mu = \{\mu_r \geq 0\}_{r=1}^k$ with $\mu_1 > 0$, $\mathscr{K}^x_{t_f,\mathscr{H}_{11}^f,\mathscr{H}_{12}^f,\mathscr{D}_{11}^f,\mathscr{D}_{21}^f,\mathscr{H}_{22}^f,\mathscr{D}^f;\mu}$, $\mathscr{K}^{z_1}_{t_f,\mathscr{H}_{11}^f,\mathscr{H}_{12}^f,\mathscr{H}_{22}^f,\mathscr{D}_{11}^f,\mathscr{D}_{21}^f,\mathscr{D}^f;\mu}$ and $\mathscr{V}_{t_f,\mathscr{H}_{11}^f,\mathscr{H}_{12}^f,\mathscr{H}_{22}^f,\mathscr{D}_{11}^f,\mathscr{D}_{21}^f,\mathscr{D}^f;\mu}$ are the classes of $\mathscr{C}([t_0,t_f];\mathbb{R}^{m\times n})$, $\mathscr{C}([t_0,t_f];\mathbb{R}^{m\times n_1})$ and $\mathscr{C}([t_0,t_f];\mathbb{R}^m)$ with values $K_x(\cdot) \in \overline{K}_x$, $K_{z_1}(\cdot) \in \overline{K}_{z_1}$, and $v(\cdot) \in \overline{V}$ for which unique and bounded solutions to the dynamical equations (5.37)–(5.42) exist on the finite horizon $[t_0,t_f]$.

Next optimization statements for risk-averse control of the model-following problem over a finite horizon are stated.

Definition 5.3.3 (Optimization Problem of Mayer Type). Suppose $k \in \mathbb{Z}^+$ and the sequence $\mu = \{\mu_r \geq 0\}_{r=1}^k$ with $\mu_1 > 0$ are fixed. Then, the optimization problem for risk-averse control of model-following systems is defined by the minimization of value and risk-aware performance index (5.43) over the admissible sets $\mathscr{K}^x_{t_f,\mathscr{H}_{11}^f,\mathscr{H}_{12}^f,\mathscr{H}_{22}^f,\mathscr{D}_{11}^f,\mathscr{D}_{21}^f,\mathscr{D}^f;\mu}$, $\mathscr{K}^{z_1}_{t_f,\mathscr{H}_{11}^f,\mathscr{H}_{12}^f,\mathscr{H}_{22}^f,\mathscr{D}_{11}^f,\mathscr{D}_{21}^f,\mathscr{D}^f;\mu}$, $\mathscr{V}_{t_f,\mathscr{H}_{11}^f,\mathscr{H}_{12}^f,\mathscr{H}_{22}^f,\mathscr{D}_{11}^f,\mathscr{D}_{21}^f,\mathscr{D}^f;\mu}$ and subject to the dynamical equations (5.37)–(5.42) on $[t_0,t_f]$.

Upon developing a value function associated with the dynamic programming approach, the definition is necessary.

Definition 5.3.4 (Reachable Set). Let $\mathcal{Q} \triangleq \Big\{ (\varepsilon, \mathcal{Y}_{11}, \mathcal{Y}_{12}, \mathcal{Y}_{22}, \mathcal{Z}_{11}, \mathcal{Z}_{21}, \mathcal{Z}) \in [t_0, t_f] \times (\mathbb{R}^n)^k \times (\mathbb{R}^{n\times n_1})^k \times (\mathbb{R}^{n_1})^k \times (\mathbb{R}^n)^k \times (\mathbb{R}^{n_1})^k \times \mathbb{R}^k \ni \mathcal{K}^x_{t_f, \mathcal{H}^f_{11}, \mathcal{H}^f_{12}, \mathcal{H}^f_{22}, \mathcal{D}^f_{11}, \mathcal{D}^f_{21}, \mathcal{D}^f; \mu} \times \mathcal{K}^{z_1}_{t_f, \mathcal{H}^f_{11}, \mathcal{H}^f_{12}, \mathcal{H}^f_{22}, \mathcal{D}^f_{11}, \mathcal{D}^f_{21}, \mathcal{D}^f; \mu} \times \mathcal{V}_{t_f, \mathcal{H}^f_{11}, \mathcal{H}^f_{12}, \mathcal{H}^f_{22}, \mathcal{D}^f_{11}, \mathcal{D}^f_{21}, \mathcal{D}^f; \mu} \neq \emptyset \Big\}$.

Next the Hamilton-Jacobi-Bellman (HJB) equation satisfied by the value function $\mathcal{V}(\varepsilon, \mathcal{Y}_{11}, \mathcal{Y}_{12}, \mathcal{Y}_{22}, \mathcal{Z}_{11}, \mathcal{Z}_{21}, \mathcal{Z})$ as the infimum of the performance index (5.43) over the admissible classes $\mathcal{K}^x_{\varepsilon, \mathcal{Y}_{11}, \mathcal{Y}_{12}, \mathcal{Y}_{22}, \mathcal{Z}_{11}, \mathcal{Z}_{21}, \mathcal{Z}; \mu}$, $\mathcal{K}^{z_1}_{\varepsilon, \mathcal{Y}_{11}, \mathcal{Y}^f_{12}, \mathcal{Y}_{22}, \mathcal{Z}^f_{11}, \mathcal{Z}^f_{21}, \mathcal{Z}^f; \mu}$, and $\mathcal{V}_{\varepsilon, \mathcal{Y}_{11}, \mathcal{Y}_{12}, \mathcal{Y}_{22}, \mathcal{Z}_{11}, \mathcal{Z}_{21}, \mathcal{Z}; \mu}$ is now described as below.

Theorem 5.3.1 (HJB Equation for Mayer Problem). *Let $(\varepsilon, \mathcal{Y}_{11}, \mathcal{Y}_{12}, \mathcal{Y}_{22}, \mathcal{Z}_{11}, \mathcal{Z}_{21}, \mathcal{Z})$ be any interior point of the reachable set $\mathcal{Q}$ at which the value function $\mathcal{V}(\varepsilon, \mathcal{Y}_{11}, \mathcal{Y}_{12}, \mathcal{Y}_{22}, \mathcal{Z}_{11}, \mathcal{Z}_{21}, \mathcal{Z})$ is differentiable. If there exist optimal feedback $K_x^* \in \mathcal{K}^x_{t_f, \mathcal{H}^f_{11}, \mathcal{H}^f_{12}, \mathcal{H}^f_{22}, \mathcal{D}^f_{11}, \mathcal{D}^f_{21}, \mathcal{D}^f; \mu}$, optimal feedforward $K_{z_1}^* \in \mathcal{K}^{z_1}_{t_f, \mathcal{H}^f_{11}, \mathcal{H}^f_{12}, \mathcal{H}^f_{22}, \mathcal{D}^f_{11}, \mathcal{D}^f_{21}, \mathcal{D}^f; \mu}$ and optimal affine $v^* \in \mathcal{V}_{t_f, \mathcal{H}^f_{11}, \mathcal{H}^f_{12}, \mathcal{H}^f_{22}, \mathcal{D}^f_{11}, \mathcal{D}^f_{21}, \mathcal{D}^f; \mu}$, then the partial differential equation of dynamic programming*

$$
\begin{aligned}
0 = \min_{P_x \in \overline{K}_x, K_{z_1} \in \overline{K}_{z_1}, v \in \overline{V}} \Big\{ & \frac{\partial}{\partial \varepsilon} \mathcal{V}(\varepsilon, \mathcal{Y}_{11}, \mathcal{Y}_{12}, \mathcal{Y}_{22}, \mathcal{Z}_{11}, \mathcal{Z}_{21}, \mathcal{Z}) \\
& + \frac{\partial}{\partial \operatorname{vec}(\mathcal{Y}_{11})} \mathcal{V}(\varepsilon, \mathcal{Y}_{11}, \mathcal{Y}_{12}, \mathcal{Y}_{22}, \mathcal{Z}_{11}, \mathcal{Z}_{21}, \mathcal{Z}) \operatorname{vec}(\mathcal{F}_{11}(\varepsilon, \mathcal{Y}_{11}, K_x)) \\
& + \frac{\partial}{\partial \operatorname{vec}(\mathcal{Y}_{12})} \mathcal{V}(\varepsilon, \mathcal{Y}_{11}, \mathcal{Y}_{12}, \mathcal{Y}_{22}, \mathcal{Z}_{11}, \mathcal{Z}_{21}, \mathcal{Z}) \operatorname{vec}(\mathcal{F}_{12}(\varepsilon, \mathcal{Y}_{12}, \mathcal{Y}_{11}, K_x, K_{z_1})) \\
& + \frac{\partial}{\partial \operatorname{vec}(\mathcal{Y}_{22})} \mathcal{V}(\varepsilon, \mathcal{Y}_{11}, \mathcal{Y}_{12}, \mathcal{Y}_{22}, \mathcal{Z}_{11}, \mathcal{Z}_{21}, \mathcal{Z}) \operatorname{vec}(\mathcal{F}_{22}(\varepsilon, \mathcal{Y}_{22}, \mathcal{Y}_{12}, K_{z_1})) \\
& + \frac{\partial}{\partial \operatorname{vec}(\mathcal{Z}_{11})} \mathcal{V}(\varepsilon, \mathcal{Y}_{11}, \mathcal{Y}_{12}, \mathcal{Y}_{22}, \mathcal{Z}_{11}, \mathcal{Z}_{21}, \mathcal{Z}) \operatorname{vec}(\mathcal{G}_{11}(\varepsilon, \mathcal{Z}_{11}, \mathcal{Y}_{11}, \mathcal{Y}_{12}, K_x, v)) \\
& + \frac{\partial}{\partial \operatorname{vec}(\mathcal{Z}_{21})} \mathcal{V}(\varepsilon, \mathcal{Y}_{11}, \mathcal{Y}_{12}, \mathcal{Y}_{22}, \mathcal{Z}_{11}, \mathcal{Z}_{21}, \mathcal{Z}) \operatorname{vec}(\mathcal{G}_{21}(\varepsilon, \mathcal{Z}_{21}, \mathcal{Z}_{11}, \mathcal{Y}_{12}, \mathcal{Y}_{22}, K_{z_1}, v)) \\
& + \frac{\partial}{\partial \operatorname{vec}(\mathcal{Z})} \mathcal{V}(\varepsilon, \mathcal{Y}_{11}, \mathcal{Y}_{12}, \mathcal{Y}_{22}, \mathcal{Z}_{11}, \mathcal{Z}_{21}, \mathcal{Z}) \operatorname{vec}(\mathcal{G}(\varepsilon, \mathcal{Y}_{11}, \mathcal{Z}_{11}, \mathcal{Z}_{21}, v)) \Big\}
\end{aligned} \tag{5.44}
$$

is satisfied. The boundary condition is

$$
\begin{aligned}
&\mathcal{V}(t_0, \mathcal{Y}_{11}(t_0), \mathcal{Y}_{12}(t_0), \mathcal{Y}_{22}(t_0), \mathcal{Z}_{11}(t_0), \mathcal{Z}_{21}(t_0), \mathcal{Z}(t_0)) \\
&= \phi_0^m(t_0, \mathcal{H}_{11}(t_0), \mathcal{H}_{12}(t_0), \mathcal{H}_{22}(t_0), \mathcal{D}_{11}(t_0), \mathcal{D}_{21}(t_0), \mathcal{D}(t_0)).
\end{aligned}
$$

Proof. Interested readers are now referred to the work by the author available at [7]. □

Also, the reachable set $\mathcal{Q}$ contains a set of points $(\varepsilon, \mathcal{Y}_{11}, \mathcal{Y}_{12}, \mathcal{Y}_{22}, \mathcal{Z}_{11}, \mathcal{Z}_{21}, \mathcal{Z})$ from which it is possible to reach the target set $\mathcal{M}$ that is a closed subset of $[t_0, t_f] \times (\mathbb{R}^n)^k \times (\mathbb{R}^{n\times n_1})^k \times (\mathbb{R}^{n_1})^k \times (\mathbb{R}^n)^k \times (\mathbb{R}^{n_1})^k \times \mathbb{R}^k$ with some trajectories corresponding to admissible feedback gain K_x, feedforward gain K_{z_1}, and affine input v. In view of $\mathscr{Q}$, there exist some candidate functions for the value function which can be actually constructed as illustrated below.

Theorem 5.3.2 (Verification Theorem). *Fix $k \in \mathbb{Z}^+$ and let $\mathscr{W}(\varepsilon, \mathscr{Y}_{11}, \mathscr{Y}_{12}, \mathscr{Y}_{22}, \mathscr{Z}_{11}, \mathscr{Z}_{21}, \mathscr{Z})$ be a continuously differentiable solution of the HJB equation (5.44) which satisfies the boundary condition*

$$\mathscr{W}(\varepsilon, \mathscr{Y}_{11}, \mathscr{Y}_{12}, \mathscr{Y}_{22}, \mathscr{Z}_{11}, \mathscr{Z}_{21}, \mathscr{Z}) = \phi_0^m(\varepsilon, \mathscr{Y}_{11}, \mathscr{Y}_{12}, \mathscr{Y}_{22}, \mathscr{Z}_{11}, \mathscr{Z}_{21}, \mathscr{Z})$$

provided that $(\varepsilon, \mathscr{Y}_{11}, \mathscr{Y}_{12}, \mathscr{Y}_{22}, \mathscr{Z}_{11}, \mathscr{Z}_{21}, \mathscr{Z}) \in \mathscr{M}$.

Let $(t_f, \mathscr{H}_{11}^f, \mathscr{H}_{12}^f, \mathscr{H}_{22}^f, \mathscr{D}_{11}^f, \mathscr{D}_{21}^f, \mathscr{D}^f)$ be in the reachable set $\mathscr{Q}$; (K_x, K_{z_1}, v) be in $\mathscr{K}^x_{t_f,\mathscr{H}^f_{11},\mathscr{H}^f_{12},\mathscr{H}^f_{22},\mathscr{D}^f_{11},\mathscr{D}^f_{21},\mathscr{D}^f;\mu} \times \mathscr{K}^{z_1}_{t_f,\mathscr{H}^f_{11},\mathscr{H}^f_{12},\mathscr{H}^f_{22},\mathscr{D}^f_{11},\mathscr{D}^f_{21},\mathscr{D}^f;\mu} \times \mathscr{V}_{t_f,\mathscr{H}^f_{11},\mathscr{H}^f_{12},\mathscr{H}^f_{22},\mathscr{D}^f_{11},\mathscr{D}^f_{21},\mathscr{D}^f;\mu}$; $\mathscr{H}_{11}$, $\mathscr{H}_{12}$, $\mathscr{H}_{22}$, $\mathscr{D}_{11}$, $\mathscr{D}_{21}$; and $\mathscr{D}$ the corresponding solutions of the dynamical equations (5.37)–(5.42). Then, $\mathscr{W}(\tau, \mathscr{H}_{11}(\tau), \mathscr{H}_{12}(\tau), \mathscr{H}_{22}(\tau), \mathscr{D}_{11}(\tau), \mathscr{D}_{21}(\tau), \mathscr{D}(\tau))$ is a nonincreasing function of τ.

If the 3-tuple $(K_x^*, K_{z_1}^*, v^*)$ is in admissible classes $\mathscr{K}^x_{t_f,\mathscr{H}^f_{11},\mathscr{H}^f_{12},\mathscr{H}^f_{22},\mathscr{D}^f_{11},\mathscr{D}^f_{21},\mathscr{D}^f;\mu} \times \mathscr{K}^{z_1}_{t_f,\mathscr{H}^f_{11},\mathscr{H}^f_{12},\mathscr{H}^f_{22},\mathscr{D}^f_{11},\mathscr{D}^f_{21},\mathscr{D}^f;\mu} \times \mathscr{V}_{t_f,\mathscr{H}^f_{11},\mathscr{H}^f_{12},\mathscr{H}^f_{22},\mathscr{D}^f_{11},\mathscr{D}^f_{21},\mathscr{D}^f;\mu}$ defined on $[t_0, t_f]$ with the corresponding solutions, $\mathscr{H}_{11}^*$, $\mathscr{H}_{12}^*$, $\mathscr{H}_{22}^*$, $\mathscr{D}_{11}^*$, $\mathscr{D}_{21}^*$ and $\mathscr{D}^*$ of the dynamical equations (5.37)–(5.42) such that

$$\begin{aligned}
0 &= \frac{\partial}{\partial \tau}\mathscr{W}(\tau, \mathscr{H}_{11}^*(\tau), \mathscr{H}_{12}^*(\tau), \mathscr{H}_{22}^*(\tau), \mathscr{D}_{11}^*(\tau), \mathscr{D}_{21}^*(\tau), \mathscr{D}^*(\tau)) \\
&\quad + \frac{\partial}{\partial \operatorname{vec}(\mathscr{Y}_{11})}\mathscr{W}(\tau, \mathscr{H}_{11}^*(\tau), \mathscr{H}_{12}^*(\tau), \mathscr{H}_{22}^*(\tau), \mathscr{D}_{11}^*(\tau), \mathscr{D}_{21}^*(\tau), \mathscr{D}^*(\tau)) \\
&\quad \times \operatorname{vec}(\mathscr{F}_{11}(\tau, \mathscr{H}_{11}^*(\tau), K_x^*(\tau))) \\
&\quad + \frac{\partial}{\partial \operatorname{vec}(\mathscr{Y}_{12})}\mathscr{W}(\tau, \mathscr{H}_{11}^*(\tau), \mathscr{H}_{12}^*(\tau), \mathscr{H}_{22}^*(\tau), \mathscr{D}_{11}^*(\tau), \mathscr{D}_{21}^*(\tau), \mathscr{D}^*(\tau)) \\
&\quad \times \operatorname{vec}(\mathscr{F}_{12}(\tau, \mathscr{H}_{12}^*(\tau), \mathscr{H}_{11}^*(\tau), K_x^*(\tau), K_{z_1}^*(\tau))) \\
&\quad + \frac{\partial}{\partial \operatorname{vec}(\mathscr{Y}_{22})}\mathscr{W}(\tau, \mathscr{H}_{11}^*(\tau), \mathscr{H}_{12}^*(\tau), \mathscr{H}_{22}^*(\tau), \mathscr{D}_{11}^*(\tau), \mathscr{D}_{21}^*(\tau), \mathscr{D}^*(\tau)) \\
&\quad \times \operatorname{vec}(\mathscr{F}_{22}(\tau, \mathscr{H}_{22}^*(\tau), \mathscr{H}_{12}^*(\tau), K_{z_1}^*(\tau))) \\
&\quad + \frac{\partial}{\partial \operatorname{vec}(\mathscr{Z}_{11})}\mathscr{W}(\tau, \mathscr{H}_{11}^*(\tau), \mathscr{H}_{12}^*(\tau), \mathscr{H}_{22}^*(\tau), \mathscr{Z}_{11}^*(\tau), \mathscr{Z}_{21}^*(\tau), \mathscr{D}^*(\tau)) \\
&\quad \times \operatorname{vec}(\mathscr{G}_{11}(\tau, \mathscr{D}_{11}^*(\tau), \mathscr{H}_{11}^*(\tau), \mathscr{H}_{12}^*(\tau), K_x^*(\tau), v^*(\tau))) \\
&\quad + \frac{\partial}{\partial \operatorname{vec}(\mathscr{Z}_{21})}\mathscr{W}(\tau, \mathscr{H}_{11}^*(\tau), \mathscr{H}_{12}^*(\tau), \mathscr{H}_{22}^*(\tau), \mathscr{D}_{11}^*(\tau), \mathscr{D}_{21}^*(\tau), \mathscr{D}^*(\tau))
\end{aligned}$$

$$\times \operatorname{vec}(\mathscr{G}_{21}(\tau,\mathscr{D}_{21}^*(\tau),\mathscr{D}_{11}^*(\tau),\mathscr{H}_{12}^*(\tau),\mathscr{H}_{22}^*(\tau),K_{z_1}^*(\tau),v^*(\tau)))$$
$$+\frac{\partial}{\partial \operatorname{vec}(\mathscr{Z})}\mathscr{W}(\tau,\mathscr{H}_{11}^*(\tau),\mathscr{H}_{12}^*(\tau),\mathscr{H}_{22}^*(\tau),\mathscr{D}_{11}^*(\tau),\mathscr{D}_{21}^*(\tau),\mathscr{D}^*(\tau))$$
$$\times \operatorname{vec}(\mathscr{G}(\tau,\mathscr{H}_{11}^*(\tau),\mathscr{D}_{11}^*(\tau),\mathscr{D}_{21}^*(\tau),v^*(\tau))) \tag{5.45}$$

then the elements of K_x^*, $K_{z_1}^*$, *and* v^* *are optimal. Moreover, there holds*

$$\mathscr{W}(\varepsilon,\mathscr{Y}_{11},\mathscr{Y}_{12},\mathscr{Y}_{22},\mathscr{Z}_{11},\mathscr{Z}_{21},\mathscr{Z})=\mathscr{V}(\varepsilon,\mathscr{Y}_{11},\mathscr{Y}_{12},\mathscr{Y}_{22},\mathscr{Z}_{11},\mathscr{Z}_{21},\mathscr{Z})$$

whereby the value function is $\mathscr{V}(\varepsilon,\mathscr{Y}_{11},\mathscr{Y}_{12},\mathscr{Y}_{22},\mathscr{Z}_{11},\mathscr{Z}_{21},\mathscr{Z})$.

Proof. Mathematical analysis for the proof can be found from the previous work available at [7]. □

5.4 Existence of Risk-Averse Control Solution

At this point it is worthwhile to note that the optimization problem herein can be solved by applying an adaptation of the Mayer-form verification theorem in [8]. This in turn implies that the terminal-value condition $(t_f,\mathscr{H}_{11}^f,\mathscr{H}_{12}^f,\mathscr{H}_{22}^f,\mathscr{D}_{11}^f,\mathscr{D}_{21}^f,\mathscr{D}^f)$ of the dynamics (5.37)–(5.42) are parameterized as $(\varepsilon,\mathscr{Y}_{11},\mathscr{Y}_{12},\mathscr{Y}_{22},\mathscr{Z}_{11},\mathscr{Z}_{21},\mathscr{Z})$ for a broader family of optimization problems. With the background provided by the representation of the risk-value aware performance index (5.43), the choice of a real-valued function considered as a candidate solution to the HJB equation (5.44) is given by

$$\mathscr{W}(\varepsilon,\mathscr{Y}_{11},\mathscr{Y}_{12},\mathscr{Y}_{22},\mathscr{Z}_{11},\mathscr{Z}_{21},\mathscr{Z})=x_0^T\sum_{r=1}^k\mu_r(\mathscr{Y}_{11}^r+\mathscr{E}_{11}^r(\varepsilon))x_0$$
$$+2x_0^T\sum_{r=1}^k\mu_r(\mathscr{Y}_{12}^r+\mathscr{E}_{12}^r(\varepsilon))z_{10}+z_{10}^T\sum_{r=1}^k\mu_r(\mathscr{Y}_{22}^r+\mathscr{E}_{22}^r(\varepsilon))z_{10}+\sum_{r=1}^k\mu_r(\mathscr{Z}^r+\mathscr{T}^r(\varepsilon))$$
$$+2x_0^T\sum_{r=1}^k\mu_r(\mathscr{Z}_{11}^r+\mathscr{T}_{11}^r(\varepsilon))+2z_{10}^T\sum_{r=1}^k\mu_r(\mathscr{Z}_{21}^r+\mathscr{T}_{21}^r(\varepsilon)), \tag{5.46}$$

whereby $(\varepsilon,\mathscr{Y}_{11},\mathscr{Y}_{12},\mathscr{Y}_{22},\mathscr{Z}_{11},\mathscr{Z}_{21},\mathscr{Z})$ is any interior point of the reachable set $\mathscr{Q}$ while the time-parametric functions $\mathscr{E}_{11}^r\in\mathscr{C}^1([t_0,t_f];\mathbb{R}^n)$, $\mathscr{E}_{12}^r\in\mathscr{C}^1([t_0,t_f];\mathbb{R}^{n\times n_1})$, $\mathscr{E}_{22}^r\in\mathscr{C}^1([t_0,t_f];\mathbb{R}^{n_1})$, $\mathscr{T}_{11}^r\in\mathscr{C}^1([t_0,t_f];\mathbb{R}^n)$, $\mathscr{T}_{21}^r\in\mathscr{C}^1([t_0,t_f];\mathbb{R}^{n_1})$ and $\mathscr{T}^r\in\mathscr{C}^1([t_0,t_f];\mathbb{R})$ are to be determined.

In addition, it is straightforward to show that the derivative of the candidate $\mathscr{W}(\varepsilon, \mathscr{Y}_{11}, \mathscr{Y}_{12}, \mathscr{Y}_{22}, \mathscr{Z}_{11}, \mathscr{Z}_{21}, \mathscr{Z})$ for the value function with respect to time ε is

$$\frac{d}{d\varepsilon}\mathscr{W}(\varepsilon, \mathscr{Y}_{11}, \mathscr{Y}_{12}, \mathscr{Y}_{22}, \mathscr{Z}_{11}, \mathscr{Z}_{21}, \mathscr{Z}) = x_0^T \sum_{r=1}^{k} \mu_r(\mathscr{F}_{11}^r(\varepsilon, \mathscr{Y}_{11}, K_x) + \frac{d}{d\varepsilon}\mathscr{E}_{11}^r(\varepsilon))x_0$$
$$+ 2x_0^T \sum_{r=1}^{k} \mu_r(\mathscr{F}_{12}^r(\varepsilon, \mathscr{Y}_{12}, \mathscr{Y}_{11}, K_x, K_{z_1}) + \frac{d}{d\varepsilon}\mathscr{E}_{12}^r(\varepsilon))z_{10}$$
$$+ z_{10}^T \sum_{r=1}^{k} \mu_r(\mathscr{F}_{22}^r(\varepsilon, \mathscr{Y}_{22}, \mathscr{Y}_{12}, K_{z_1}) + \frac{d}{d\varepsilon}\mathscr{E}_{22}^r(\varepsilon))z_{10}$$
$$+ 2x_0^T \sum_{r=1}^{k} \mu_r(\mathscr{G}_{11}^r(\varepsilon, \mathscr{Z}_{11}, \mathscr{Y}_{11}, \mathscr{Y}_{12}, K_x, v) + \frac{d}{d\varepsilon}\mathscr{T}_{11}^r(\varepsilon))$$
$$+ 2z_{10}^T \sum_{r=1}^{k} \mu_r(\mathscr{G}_{21}^r(\varepsilon, \mathscr{Z}_{21}, \mathscr{Z}_{11}, \mathscr{Y}_{12}, \mathscr{Y}_{22}, K_{z_1}, v) + \frac{d}{d\varepsilon}\mathscr{T}_{21}^r(\varepsilon))$$
$$+ \sum_{r=1}^{k} \mu_r(\mathscr{G}^r(\varepsilon, \mathscr{Y}_{11}, \mathscr{Z}_{11}, \mathscr{Z}_{21}, v) + \frac{d}{d\varepsilon}\mathscr{T}^r(\varepsilon)) \qquad (5.47)$$

provided that $(K_x, K_{z_1}, v) \in \overline{P}_x \times \overline{P}_{z_1} \times \overline{V}$.

The substitution of this candidate for the value function into the HJB equation (5.44) and making use of Eq. (5.47) yield

$$0 = \min_{(K_x, K_{z_1}, v) \in \overline{P}_x \times \overline{P}_{z_1} \times \overline{V}} \Bigg\{ x_0^T \sum_{r=1}^{k} \mu_r(\mathscr{F}_{11}^r(\varepsilon, \mathscr{Y}_{11}, K_x) + \frac{d}{d\varepsilon}\mathscr{E}_{11}^r(\varepsilon))x_0$$
$$+ 2x_0^T \sum_{r=1}^{k} \mu_r(\mathscr{F}_{12}^r(\varepsilon, \mathscr{Y}_{12}, \mathscr{Y}_{11}, K_x, K_{z_1}) + \frac{d}{d\varepsilon}\mathscr{E}_{12}^r(\varepsilon))z_{10}$$
$$+ z_{10}^T \sum_{r=1}^{k} \mu_r(\mathscr{F}_{22}^r(\varepsilon, \mathscr{Y}_{22}, \mathscr{Y}_{12}, K_{z_1}) + \frac{d}{d\varepsilon}\mathscr{E}_{22}^r(\varepsilon))z_{10}$$
$$+ 2x_0^T \sum_{r=1}^{k} \mu_r(\mathscr{G}_{11}^r(\varepsilon, \mathscr{Z}_{11}, \mathscr{Y}_{11}, \mathscr{Y}_{12}, K_x, v) + \frac{d}{d\varepsilon}\mathscr{T}_{11}^r(\varepsilon))$$
$$+ 2z_{10}^T \sum_{r=1}^{k} \mu_r(\mathscr{G}_{21}^r(\varepsilon, \mathscr{Z}_{21}, \mathscr{Z}_{11}, \mathscr{Y}_{12}, \mathscr{Y}_{22}, K_{z_1}, v) + \frac{d}{d\varepsilon}\mathscr{T}_{21}^r(\varepsilon))$$
$$+ \sum_{r=1}^{k} \mu_r(\mathscr{G}^r(\varepsilon, \mathscr{Y}_{11}, \mathscr{Z}_{11}, \mathscr{Z}_{21}, v) + \frac{d}{d\varepsilon}\mathscr{T}^r(\varepsilon)) \Bigg\}. \qquad (5.48)$$

Differentiating the expression within the bracket of Eq. (5.48) with respect to K_x, K_{z_1}, and v yield the necessary conditions for an extremum of the Mayer-type performance index (5.43) on the time interval $[t_0, \varepsilon]$

$$K_x(\varepsilon, \mathscr{Y}_{11}) = -R^{-1}(\varepsilon)B^T(\varepsilon)\sum_{r=1}^{k} \hat{\mu}_r \mathscr{Y}_{11}^r, \tag{5.49}$$

$$K_{z_1}(\varepsilon, \mathscr{Y}_{12}) = -R^{-1}(\varepsilon)B^T(\varepsilon)\sum_{r=1}^{k} \hat{\mu}_r \mathscr{Y}_{12}^r, \tag{5.50}$$

$$v(\varepsilon, \mathscr{Z}_{11}) = -R^{-1}(\varepsilon)B^T(\varepsilon)\sum_{r=1}^{k} \hat{\mu}_r \mathscr{Z}_{11}^r, \qquad \hat{\mu}_r = \mu_r/\mu_1. \tag{5.51}$$

Furthermore, the dynamical equations (5.37)–(5.42) evaluated at the feedback (5.49), the feedforward (5.50), and the affine input (5.51) become the backward-in-time matrix matrix-valued differential equations

$$\frac{d}{d\varepsilon}\mathscr{H}_{11}^1(\varepsilon) = -[A(\varepsilon)+B(\varepsilon)K_x(\varepsilon)]^T\mathscr{H}_{11}^1(\varepsilon) - \mathscr{H}_{11}^1(\varepsilon)[A(\varepsilon)+B(\varepsilon)K_x(\varepsilon)]$$
$$-C^T(\varepsilon)Q(\varepsilon)C(\varepsilon) - K_x^T(\varepsilon)R(\varepsilon)K_x(\varepsilon), \tag{5.52}$$

$$\frac{d}{d\varepsilon}\mathscr{H}_{11}^r(\varepsilon) = -[A(\varepsilon)+B(\varepsilon)K_x(\varepsilon)]^T\mathscr{H}_{11}^r(\varepsilon) - \mathscr{H}_{11}^r(\varepsilon)[A(\varepsilon)+B(\varepsilon)K_x(\varepsilon)]$$
$$-\sum_{s=1}^{r-1}\frac{2r!}{s!(r-s)!}\mathscr{H}_{11}^s(\varepsilon)G(\varepsilon)WG^T(\varepsilon)\mathscr{H}_{11}^{(r-s)}(\varepsilon), \tag{5.53}$$

$$\frac{d}{d\varepsilon}\mathscr{H}_{12}^1(\varepsilon) = -[A(\varepsilon)+B(\varepsilon)K_x(\varepsilon)]^T\mathscr{H}_{12}^1(\varepsilon) - \mathscr{H}_{11}^1(\varepsilon)B(\varepsilon)K_{z_1}(\varepsilon) - \mathscr{H}_{12}^1(\varepsilon)A_1(\varepsilon)$$
$$+C^T(\varepsilon)Q(\varepsilon)C_1(\varepsilon) - K_x^T(\varepsilon)R(\varepsilon)K_{z_1}(\varepsilon), \tag{5.54}$$

$$\frac{d}{d\varepsilon}\mathscr{H}_{12}^r(\varepsilon) = -[A(\varepsilon)+B(\varepsilon)K_x(\varepsilon)]^T\mathscr{H}_{12}^r(\varepsilon) - \mathscr{H}_{11}^r(\varepsilon)B(\varepsilon)K_{z_1}(\varepsilon) - \mathscr{H}_{12}^r(\varepsilon)A_1(\varepsilon)$$
$$-\sum_{s=1}^{r-1}\frac{2r!}{s!(r-s)!}\mathscr{H}_{11}^s(\varepsilon)G(\tau)WG^T(\varepsilon)\mathscr{H}_{12}^{(r-s)}(\varepsilon), \tag{5.55}$$

$$\frac{d}{d\varepsilon}\mathscr{H}_{22}^1(\varepsilon) = -A_1^T(\varepsilon)\mathscr{H}_{22}^1(\varepsilon) - \mathscr{H}_{22}^1(\varepsilon)A_1(\varepsilon) - K_{z_1}^T(\varepsilon)R(\varepsilon)K_{z_1}(\varepsilon)$$
$$-K_{z_1}^T(\varepsilon)B^T(\varepsilon)\mathscr{H}_{12}^1(\varepsilon) - (\mathscr{H}_{12}^1)^T(\varepsilon)B(\varepsilon)K_{z_1}(\varepsilon) - C_1^T(\varepsilon)Q(\varepsilon)C_1(\varepsilon) \tag{5.56}$$

$$\frac{d}{d\varepsilon}\mathscr{H}_{22}^r(\varepsilon) = -A_1^T(\varepsilon)\mathscr{H}_{22}^r(\varepsilon) - (\mathscr{H}_{12}^r)^T(\varepsilon)B(\varepsilon)K_{z_1}(\varepsilon) - K_{z_1}^T(\varepsilon)B^T(\varepsilon)\mathscr{H}_{12}^r(\varepsilon)$$
$$-\mathscr{H}_{22}^r(\varepsilon)A_1(\varepsilon) - \sum_{s=1}^{r-1}\frac{2r!}{s!(r-s)!}(\mathscr{H}_{12}^s)^T(\varepsilon)G(\varepsilon)WG^T(\varepsilon)\mathscr{H}_{12}^{r-s}(\varepsilon) \tag{5.57}$$

the backward-in-time vector-valued differential equations

$$\frac{d}{d\varepsilon}\mathcal{D}_{11}^1(\varepsilon) = -\left[A(\varepsilon)+B(\tau)K_x(\varepsilon)\right]^T \mathcal{D}_{11}^1(\varepsilon) - \mathcal{H}_{11}^1(\varepsilon)B(\varepsilon)v(\varepsilon) - \mathcal{H}_{12}^1(\varepsilon)B_1(\varepsilon)u_r(\varepsilon) - K_x^T(\varepsilon)R(\varepsilon)v(\varepsilon), \quad (5.58)$$

$$\frac{d}{d\varepsilon}\mathcal{D}_{11}^r(\varepsilon) = -\left[A(\varepsilon)+B(\varepsilon)K_x(\varepsilon)\right]^T \mathcal{D}_{11}^r(\varepsilon) - \mathcal{H}_{11}^r(\varepsilon)B(\varepsilon)v(\varepsilon) - \mathcal{H}_{12}^r(\varepsilon)B_1(\varepsilon)u_r(\varepsilon), \quad (5.59)$$

$$\frac{d}{d\varepsilon}\mathcal{D}_{21}^1(\varepsilon) = -A_1^T(\varepsilon)\mathcal{D}_{21}^1(\varepsilon) - K_{z_1}^T(\varepsilon)B^T(\varepsilon)\mathcal{D}_{11}^1(\varepsilon) - (\mathcal{H}_{12}^1)^T(\varepsilon)B(\varepsilon)v(\varepsilon) - \mathcal{H}_{22}^1(\varepsilon)B_1(\varepsilon)u_r(\varepsilon) - K_{z_1}^T(\varepsilon)R(\varepsilon)v(\varepsilon), \quad (5.60)$$

$$\frac{d}{d\varepsilon}\mathcal{D}_{21}^r(\varepsilon) = -A_1^T(\varepsilon)\mathcal{D}_{21}^r(\varepsilon) - K_{z_1}^T(\varepsilon)B^T(\varepsilon)\mathcal{D}_{11}^r(\varepsilon) - (\mathcal{H}_{12}^r)^T(\varepsilon)B(\tau)v(\varepsilon) - \mathcal{H}_{22}^r(\varepsilon)B_1(\varepsilon)u_r(\varepsilon) \quad (5.61)$$

and the backward-in-time scalar-valued differential equations

$$\frac{d}{d\varepsilon}\mathcal{D}^1(\varepsilon) = -\mathrm{Tr}\left\{\mathcal{H}_{11}^1(\varepsilon)G(\varepsilon)WG^T(\varepsilon)\right\} - v^T(\varepsilon)R(\varepsilon)v(\varepsilon) - 2(\mathcal{D}_{11}^1)^T(\varepsilon)B(\varepsilon)v(\varepsilon) - 2(\mathcal{D}_{21}^1)^T(\varepsilon)B_1(\varepsilon)u_r(\varepsilon), \quad (5.62)$$

$$\frac{d}{d\varepsilon}\mathcal{D}^r(\varepsilon) = -\mathrm{Tr}\left\{\mathcal{H}_{11}^r(\varepsilon)G(\varepsilon)WG^T(\varepsilon)\right\} - 2(\mathcal{D}_{11}^r)^T(\varepsilon)B(\varepsilon)v(\varepsilon) - 2(\mathcal{D}_{21}^r)^T(\varepsilon)B_1(\varepsilon)u_r(\varepsilon), \quad (5.63)$$

whereby the terminal-value conditions are $\mathcal{H}_{11}^1(t_f) = C^T(t_f)Q_fC(t_f)$, $\mathcal{H}_{11}^r(t_f) = 0$ for $2 \le r \le k$; $\mathcal{H}_{12}^1(t_f) = -C^T(t_f)Q_fC_1(t_f)$, $\mathcal{H}_{12}^r(t_f) = 0$ for $2 \le r \le k$; $\mathcal{H}_{22}^1(t_f) = C_1^T(t_f)Q_fC_1(t_f)$, $\mathcal{H}_{22}^r(t_f) = 0$ for $2 \le r \le k$; $\mathcal{D}_{11}^r(t_f) = 0$ for $1 \le r \le k$; $\mathcal{D}_{21}^r(t_f) = 0$ for $1 \le r \le k$; $\mathcal{D}^r(t_f) = 0$ for $1 \le r \le k$.

To have the minimum of Eq. (5.48) equal to zero for any $\varepsilon \in [t_0, t_f]$ and when $\mathcal{Y}_{11}^r$, $\mathcal{Y}_{12}^r$, $\mathcal{Y}_{22}^r$, $\mathcal{Z}_{11}^r$, $\mathcal{Z}_{21}^r$, and $\mathcal{Z}^r$ evaluated along the solutions to the equations (5.37)–(5.42), the time-dependent functions $\mathcal{E}_{11}^r(\varepsilon)$, $\mathcal{E}_{12}^r(\varepsilon)$, $\mathcal{E}_{22}^r(\varepsilon)$, $\mathcal{T}_{11}^r(\varepsilon)$, $\mathcal{T}_{21}^r(\varepsilon)$, and $\mathcal{T}^r(\varepsilon)$ must satisfy the following differential equations:

$$\frac{d}{d\varepsilon}\mathcal{E}_{11}^r(\varepsilon) = -\frac{d}{d\varepsilon}\mathcal{H}_{11}^r(\varepsilon), \qquad \frac{d}{d\varepsilon}\mathcal{E}_{12}^r(\varepsilon) = -\frac{d}{d\varepsilon}\mathcal{H}_{12}^r(\varepsilon), \quad (5.64)$$

$$\frac{d}{d\varepsilon}\mathcal{E}_{22}^r(\varepsilon) = -\frac{d}{d\varepsilon}\mathcal{H}_{22}^r(\varepsilon), \qquad \frac{d}{d\varepsilon}\mathcal{T}_{11}^r(\varepsilon) = -\frac{d}{d\varepsilon}\mathcal{D}_{11}^r(\varepsilon), \quad (5.65)$$

$$\frac{d}{d\varepsilon}\mathcal{T}_{21}^r(\varepsilon) = -\frac{d}{d\varepsilon}\mathcal{D}_{21}^r(\varepsilon), \qquad \frac{d}{d\varepsilon}\mathcal{T}^r(\varepsilon) = -\frac{d}{d\varepsilon}\mathcal{D}^r(\varepsilon). \quad (5.66)$$

According to the verification theorem for the optimization problem of Mayer type, the boundary condition of $\mathscr{W}(\varepsilon, \mathscr{Y}_{11}, \mathscr{Y}_{12}, \mathscr{Y}_{22}, \mathscr{Z}_{11}, \mathscr{Z}_{21}, \mathscr{Z})$ requires

$$
\begin{aligned}
& x_0^T \sum_{r=1}^{k} \mu_r(\mathscr{H}_{11}^r(t_0) + \mathscr{E}_{11}^r(t_0))x_0 + 2x_0^T \sum_{r=1}^{k} \mu_r(\mathscr{H}_{12}^r(t_0) + \mathscr{E}_{12}^r(t_0))z_{10} \\
& \quad + z_{10}^T \sum_{r=1}^{k} \mu_r(\mathscr{H}_{22}^r(t_0) + \mathscr{E}_{22}^r(t_0))z_{10} + 2x_0^T \sum_{r=1}^{k} \mu_r(\mathscr{D}_{11}(t_0) + \mathscr{T}_{11}^r(t_0)) \\
& \quad + 2z_{10}^T \sum_{r=1}^{k} \mu_r(\mathscr{D}_{21}^r(t_0) + \mathscr{T}_{21}^r(t_0)) + \sum_{r=1}^{k} \mu_r(\mathscr{D}^r(t_0) + \mathscr{T}^r(t_0)) \\
& = x_0^T \sum_{r=1}^{k} \mu_r \mathscr{H}_{11}^r(t_0)x_0 + 2x_0^T \sum_{r=1}^{k} \mu_r \mathscr{H}_{12}^r(t_0)z_{10} + z_{10}^T \sum_{r=1}^{k} \mu_r \mathscr{H}_{22}^r(t_0)z_{10} \\
& \quad + 2x_0^T \sum_{r=1}^{k} \mu_r \mathscr{D}_{11}(t_0) + 2z_{10}^T \sum_{r=1}^{k} \mu_r \mathscr{D}_{21}^r(t_0) + \sum_{r=1}^{k} \mu_r \mathscr{D}^r(t_0).
\end{aligned}
$$

The initial-valued conditions for the forward-in-time differential equations (5.64)–(5.66) are therefore obtained

$$
\begin{aligned}
& \mathscr{E}_{11}^r(t_0) = 0, \quad \mathscr{E}_{12}^r(t_0) = 0, \quad \mathscr{E}_{22}^r(t_0) = 0, \\
& \mathscr{T}_{11}^r(t_0) = 0, \quad \mathscr{T}_{21}^r(t_0) = 0, \quad \mathscr{T}^r(t_0) = 0, \quad 1 \le r \le k.
\end{aligned}
$$

Subsequently, the sufficient condition (5.45) of the verification theorem is finally satisfied, so the extremizing feedback gain (5.49), feedforward gain (5.50) and affine input (5.51) are optimal

$$K_x^*(\varepsilon) = -R^{-1}(\varepsilon)B^T(\varepsilon) \sum_{r=1}^{k} \hat{\mu}_r \mathscr{H}_{11}^{r*}(\varepsilon), \tag{5.67}$$

$$K_{z_1}^*(\varepsilon) = -R^{-1}(\varepsilon)B^T(\varepsilon) \sum_{r=1}^{k} \hat{\mu}_r \mathscr{H}_{12}^{r*}(\varepsilon), \tag{5.68}$$

$$v^*(\varepsilon) = -R^{-1}(\varepsilon)B^T(\varepsilon) \sum_{r=1}^{k} \hat{\mu}_r \mathscr{D}_{11}^{r*}(\varepsilon), \tag{5.69}$$

whereby the optimal matrix-valued solutions $\{\mathscr{H}_{11}^{r*}(\cdot)\}_{r=1}^k$, $\{\mathscr{H}_{12}^{r*}(\cdot)\}_{r=1}^k$, and the optimal vector-valued solutions $\{\mathscr{D}_{11}^{r*}(\cdot)\}_{r=1}^k$ are solving the backward-in-time differential equations (5.52)–(5.53), (5.54)–(5.55), and (5.58)–(5.59), respectively.

The risk-averse control design for the model-following problem over the finite horizon $[t_0, t_f]$ is shown to have the structure of three-degrees-of-freedom, e.g., $u^*(t) = K_x^*(t_f + t_0 - t)x(t) + K_{z_1}^*(t_f + t_0 - t)z_1(t) + v^*(t_f + t_0 - t)$ with the time-varying feedback gain $K_x^*(\cdot)$, the feedforward gain $K_{z_1}^*(\cdot)$, and the affine input signal $v^*(\cdot)$. The feedback gain $K_x^*(\cdot)$ of the control is independent of the system parameters of reference and command-generating models.

5.5 Chapter Summary

The new results here have demonstrated a successful attempt of integrating the compactness offered by *logic* from the *process information* (5.1)–(5.6) and the quantitativity exhibited by a priori knowledge of stationary Wiener process disturbances so that the performance information about random distributions of the goal information (5.7) can now be characterized in a compact and robust way.

Given these prerequisites on performance information and the resultant cumulants, the control process, when it is applied to the linear-quadratic class of model-following control problems, now involves three steps: (1) the establishment of standards whereby the system outputs closely track the reference outputs despite of external disturbances and initial conditions; (2) the appraisal of performance reliability is evaluated against these standards in the presence of performance risk; and (3) the correction of deviations is ensured by feedback control strategy with multiple preferences of risk aversion and performance reliability.

With regard to the use of performance information for feedback, the optimal control strategy with risk aversion has the increased complex structure of three-degrees-of-freedom, as in Eq. (5.8), within which the feedback gain (5.67) and the feedforward gain (5.68) operate dynamically on the backward-in-time matrix-valued evolutions (5.52)–(5.53) and vector-valued evolutions (5.54)–(5.55), whereas the *affine input* (5.69) is driven by the class of systems with zero input responses through the backward-in-time vector-valued evolutions (5.58)–(5.59).

References

1. Pham, K.D.: On statistical control of stochastic servo-systems: performance-measure statistics and state-feedback paradigm. In: Proceedings of the 17th International Federation of Automatic Control World Congress, pp. 7994–8000 (2008)
2. Davison, E.J.: The feedforward control of linear multivariable time-invariant systems. Automatica **9**, 561–573 (1973)
3. Davison, E.J.: The steady-state invertibility and feedforward control of linear time-invariant systems. IEEE Trans. Automat. Contr. **21**, 529–534 (1976)
4. Grimble, M.J.: Two-degrees of freedom feedback and feedforward optimal control of multivariable stochastic systems. Automatica **24**, 809–817 (1988)
5. Hassibi, B., Kailath, T.: Tracking with an H-infinity criterion. In: Proceedings of the Conference on Decision and Control, pp. 3594–3599 (1997)
6. Yuksel, S., Hindi, H., Crawford, L.: Optimal tracking with feedback-feedforward control separation over a network. In: Proceedings of the American Control Conference, pp. 3500–3506 (2006)
7. Pham, K.D.: Statistical control paradigms for structural vibration suppression. Ph.D. Dissertation. Department of Electrical Engineering. University of Notre Dame, Indiana, U.S.A. Available via http://etd.nd.edu/ETD-db/theses/available/etd-04152004-121926/unrestricted/PhamKD052004.pdf. Cited 15 March 2012 (2004)
8. Fleming, W.H., Rishel, R.W.: Deterministic and Stochastic Optimal Control. Springer, New York (1975)

Chapter 6
Incomplete Feedback Design in Model-Following Systems

Abstract The chapter extends the theory of risk-averse control of a linear-quadratic class of model-following control systems with incomplete state feedback. It is shown that performance information can improve control decisions with only available output measurements for system performance reliability. Many of the results entail measures of the amount, value, and cost of performance information, and the design of model-following control strategy with risk aversion. This thematic view of correct-by-designs of high performance and reliable systems is a double-edged sword. By using all the mathematical statistics associated with performance appraisal to treat performance reliability it provides a unique and unifying perspective of decision making with risk consequences. On the other hand, it complicates information structures of the resulting control solutions.

6.1 Introduction

As evidenced in the first four chapters, the shape and functional form of an utility function tell a great deal about the basic attitudes of the decision makers and controller designers toward the uncertain outcomes or performance risks. Of particular, the new utility function or the so-called risk-value aware performance index, which is being proposed therein as a linear manifold defined by a finite number of cost cumulants associated with a random quadratic cost functional will provide a convenient allocation representation of apportioning performance robustness and reliability requirements into the multi-attribute requirement of qualitative characteristics of expected performance and performance risks.

With regards to earlier works [1, 2, 7], recent developments [4,5] attempted to bridge the gaps between the field of tracking and feedforward controls and statistical learning whereby a class of risk-averse decisions implementing perfect state-feedback controllers pertaining to stochastic servo systems on a finite horizon. This advanced controller concept depends upon the performance information to

K.D. Pham, *Linear-Quadratic Controls in Risk-Averse Decision Making*,
SpringerBriefs in Optimization, DOI 10.1007/978-1-4614-5079-5_6,

establish a two-degrees-of-freedom description of desired feedback and feedforward actions for minimal differences between reference and system outputs and tradeoffs between risks and benefits for performance reliability.

In this chapter, some important results that are new in stochastic control of model-following systems in presence of noisy output measurements are organized as follows. Section 6.2 characterizes performance uncertainty and risk subject to underlying stochastic disturbances as applied to the problem of structuring and measuring performance variations. Section 6.3 is devoted to the problem statements in model-following systems. It shows a new paradigm of thinking with which the dynamics of cost cumulants can be incorporated into reliable control design with risk-value preferences. Section 6.4 presents a complete development of risk-averse control strategy with the three-degrees-of-freedom structure within which the system not only follows the outputs of the reference model driven by a class of command inputs but also takes into account of performance value and risk tradeoffs. Conclusions are also in Sect. 6.5.

6.2 Backward Differential Equations for Performance-Measure Statistics

Let $(\Omega, \mathscr{F}, \mathbb{F}, \mathbb{P})$ be a complete filtered probability space over $[t_0, t_f]$, on which a p-dimensional standard stationary Wiener $w(\cdot)$ with the correlation of increments $E\{[w(\tau)-w(\xi)][w(\tau)-w(\xi)]^T\} = W|\tau-\xi|, W > 0$ for all $\tau, \xi \in [t_0, t_f]$ is defined with $\mathbb{F} = \{\mathscr{F}_t\}_{t \ge t_0 \ge 0}$ being its natural filtration, augmented by all $\mathbb{P}$-null sets in $\mathscr{F}$. Next consider a class of stabilizable systems described by the *information process* which is governed by the controlled stochastic differential equation

$$dx(t) = (A(t)x(t) + B(t)u(t))\,dt + G(t)dw(t)\,, \quad x(t_0) = x_0, \tag{6.1}$$

whereby the continuous-time coefficients $A \in \mathscr{C}([t_0,t_f];\mathbb{R}^{n\times n})$, $B \in \mathscr{C}([t_0,t_f];\mathbb{R}^{n\times m})$, and $G \in \mathscr{C}([t_0,t_f];\mathbb{R}^{n\times p})$. In the above, $x(\cdot)$ is the controlled state process valued in $\mathbb{R}^n$ and $u(\cdot)$ is the control process valued in some set $U \subseteq \mathbb{R}^m$. The output involves some functions $C \in \mathscr{C}([t_0,t_f];\mathbb{R}^{r\times n})$, relating $y(\cdot)$ to the values of the manipulated process $x(\cdot)$

$$y(t) = C(t)x(t)\,. \tag{6.2}$$

In model-following control design, the desired behavior of the output (6.2) is enabled through the use of a reference model driven by a reference input. Typically, a linear model with the output reference and known state $z_1(t_0) = z_{10}$ is used

$$dz_1(t) = (A_1(t)z_1(t) + B_1(t)u_r(t))\,dt + G_1(t)dw_1(t), \tag{6.3}$$

$$y_1(t) = C_1(t)z_1(t), \tag{6.4}$$

where continuous-time coefficients $A_1 \in \mathscr{C}([t_0,t_f];\mathbb{R}^{n_1 \times n_1})$, $B_1 \in \mathscr{C}([t_0,t_f];\mathbb{R}^{n_1 \times m_1})$, $G_1 \in \mathscr{C}([t_0,t_f];\mathbb{R}^{n_1 \times p_1})$, $C_1 \in \mathscr{C}([t_0,t_f];\mathbb{R}^{r \times n_1})$ and the p_1-dimensional stationary Wiener $w_1(\cdot)$ with $E\{[w_1(\tau)-w_1(\xi)][w_1(\tau)-w_1(\xi)]^T\} = W_1|\tau-\xi|$, $W_1 > 0$ for all $\tau, \xi \in [t_0,t_f]$.

Furthermore, the reference input $u_r(\cdot)$, valued in $U_r \subseteq \mathbb{R}^{m_1}$, is also known as the command output of the class of zero-input responses of the system

$$dz_2(t) = A_2(t)z_2(t)dt\,, \qquad z_2(t_0) = z_{20}, \tag{6.5}$$

$$u_r(t) = C_2(t)z_2(t), \tag{6.6}$$

whereby some knowledge of $A_2 \in \mathscr{C}([t_0,t_f];\mathbb{R}^{n_2 \times n_2})$ and $C_2 \in \mathscr{C}([t_0,t_f];\mathbb{R}^{m_1 \times n_2})$ are required.

For each $t \in [t_0,t_f]$, the purpose of the *decision process* for the model-following control design is to use the control process $u(t)$ to force an estimate of the output process $\hat{y}(t) = C(t)\hat{x}(t)$ with accessible state $\hat{x}(t)$ to track an estimate of the desired process $\hat{y}_1(t) = C_1(t)\hat{z}_1(t)$ with accessible state $\hat{z}_1(t)$ in spite of the disturbances $w(t)$ and $w_1(t)$ as well as the initial states x_0 and z_{10}. Associated with $(t_0,x_0,z_{10};u) \in [t_0,t_f] \times \mathbb{R}^n \times \mathbb{R}^{n_1} \times U$ is a finite-horizon integral-quadratic-form (IQF) cost $J : [t_0,t_f] \times \mathbb{R}^n \times \mathbb{R}^{n_1} \times U \mapsto \mathbb{R}^+$ used for matching of the outputs

$$J(t_0,x_0,z_{10};u) = (\hat{y}(t_f)-\hat{y}_1(t_f))^T Q_f (\hat{y}(t_f)-\hat{y}_1(t_f)) + \int_{t_0}^{t_f} [(\hat{y}(\tau)-\hat{y}_1(\tau))^T Q(\tau)(\hat{y}(\tau)-\hat{y}_1(\tau)) + u^T(\tau)R(\tau)u(\tau)]d\tau \tag{6.7}$$

subject to the constraints [6]

$$d\hat{x}(t) = (A(t)\hat{x}(t) + B(t)u(t))dt + L(t)dw(t), \tag{6.8}$$

$$d\hat{z}_1(t) = (A_1(t)\hat{z}_1(t) + B_1(t)u_r(t))dt + L_1(t)dw_1(t), \tag{6.9}$$

where the initial states $\hat{x}(t_0) = x_0$ and $\hat{z}_1(t_0) = z_{10}$ known;

$$L(t) = [P(t)\tilde{C}^T(t) + WC^T(t)]\Delta^{-1}(t)C(t)G(t)\,, \quad \Delta(t) = C(t)WC^T(t), \tag{6.10}$$

$$L_1(t) = [P_1(t)\tilde{C}_1^T(t) + W_1C_1^T(t)]\Delta_1^{-1}(t)C_1(t)G_1(t),\ \Delta_1(t) = C_1(t)W_1C_1^T(t), \tag{6.11}$$

$$\tilde{C}(t) = \dot{C}(t) + C(t)A(t)\,, \quad \tilde{C}_1(t) = \dot{C}_1(t) + C_1(t)A_1(t), \tag{6.12}$$

$$\frac{d}{dt}P(t) = [A(t)-WC^T(t)\Delta^{-1}(t)\tilde{C}(t)]P(t) + P(t)[A(t)-WC(t)\Delta^{-1}(t)\tilde{C}(t)]^T + W - P(t)\tilde{C}^T(t)\Delta^{-1}(t)\tilde{C}(t)P(t) - WC^T(t)\Delta^{-1}(t)C(t)W;\ P(t_0)=0, \tag{6.13}$$

$$\frac{d}{dt}P_1(t) = [A_1(t)-W_1C_1^T(t)\Delta_1^{-1}(t)\tilde{C}_1(t)]P_1(t) + P_1(t)[A_1(t)-W_1C_1(t)\Delta_1^{-1}(t)\tilde{C}_1(t)]^T + W_1 - P_1(t)\tilde{C}_1^T(t)\Delta_1^{-1}(t)\tilde{C}_1(t)P_1(t) - W_1C_1^T(t)\Delta_1^{-1}(t)C_1(t)W_1;\ P_1(t_0)=0. \tag{6.14}$$

In addition, the terminal matching penalty $Q_f \in \mathbb{R}^r$, the matching penalty $Q \in \mathscr{C}([t_0,t_f];\mathbb{R}^r)$, and the control effort weighting $R \in \mathscr{C}([t_0,t_f];\mathbb{R}^m)$ are symmetric and positive semidefinite matrices with $R(t)$ invertible.

From Eqs. (6.13) and (6.14), the covariances of error estimates are independent of control actions and observations. Therefore, to parameterize the conditional densities $p(x(t)|\mathscr{F}_t)$ and $p(z_1(t)|\mathscr{F}_{1t})$, the conditional means $\hat{x}(t)$ and $\hat{z}_1(t)$ minimizing covariances of error estimates of $x(t)$ and $z_1(t)$ are only needed. An admissible control law is hence of the form

$$u(t) = K_x(t)\hat{x}(t) + K_{z_1}(t)\hat{z}_1(t) + v(t), \tag{6.15}$$

where $v(\cdot) \in V \subseteq \mathbb{R}^m$ is an affine control input and while the elements of $K_x \in \mathscr{C}([t_0,t_f];\mathbb{R}^{m\times n})$ and $K_{z_1} \in \mathscr{C}([t_0,t_f];\mathbb{R}^{m\times n_1})$ are admissible feedback and feedforward gains defined in appropriate senses. Hence, for the given initial condition $(t_0,x_0,z_{10},z_{20}) \in [t_0,t_f] \times \mathbb{R}^n \times \mathbb{R}^{n_1} \times \mathbb{R}^{n_2}$ and the control policy (6.15), the aggregation of Eqs. (6.8) and (6.9) is described by the controlled stochastic differential equation

$$\mathrm{d}x_a(t) = (A_a(t)x_a(t) + B_a(t)v(t) + \Gamma_a(t)u_r(t))\,\mathrm{d}t + G_a(t)\mathrm{d}w_a(t), \quad x_a(t_0) \tag{6.16}$$

with the performance measure (6.7) rewritten as

$$J(K_x(\cdot),K_{z_1}(\cdot),v(\cdot)) = x_a^T(t_f)Q_{fa}x_a(t_f) + \int_{t_0}^{t_f} [x_a^T(\tau)Q_a(\tau)x_a(\tau) + 2x_a^T(\tau)N_a(\tau)v(\tau) + v^T(\tau)R(\tau)v(\tau)]\mathrm{d}\tau, \tag{6.17}$$

whereby the augmented states $x_a^T = \begin{bmatrix} \hat{x}^T & \hat{z}_1^T \end{bmatrix}$ and continuous-time coefficients are

$$A_a = \begin{bmatrix} A+BK_x & BK_{z_1} \\ 0 & A_1 \end{bmatrix}, \ B_a = \begin{bmatrix} B \\ 0 \end{bmatrix}, \ G_a = \begin{bmatrix} L & 0 \\ 0 & L_1 \end{bmatrix}, \ x_{a0} = \begin{bmatrix} x_0 \\ z_{10} \end{bmatrix}, \ w_a = \begin{bmatrix} w \\ w_1 \end{bmatrix},$$

$$\Gamma_a = \begin{bmatrix} 0 \\ B_1 \end{bmatrix}, \ N_a = \begin{bmatrix} K_x^T R \\ K_{z_1}^T R \end{bmatrix}, \ Q_{fa} = \begin{bmatrix} C^T(t_f)Q_fC(t_f) & -C^T(t_f)Q_fC_1(t_f) \\ -C_1^T(t_f)Q_fC(t_f) & C_1^T(t_f)Q_fC_1(t_f) \end{bmatrix},$$

$$Q_a = \begin{bmatrix} C^TQC + K_x^TRK_x & -C^TQC_1 + K_x^TRK_{z_1} \\ -C_1^TQC + K_{z_1}^TRK_x & C_1^TQC_1 + K_{z_1}^TRK_{z_1} \end{bmatrix}, \quad W_a = \begin{bmatrix} W & 0 \\ 0 & W_1 \end{bmatrix}.$$

In view of two types of information, i.e., process information (6.16) and goal information (6.17) as well as the external disturbance $w_a(\cdot)$ affecting the closed-loop performance, the controller now needs additional information about performance variations. This is coupling information and thus also known as performance information. The questions of how to characterize and influence performance information are then answered by adaptive higher-order statistics associated with the performance measure (6.17) below.

Theorem 6.2.1 (Cumulant Generating Function). *Let $\varphi(\tau, x_a^\tau; \theta) \triangleq \rho_a(\tau, \theta)$ $\exp\{(x_a^\tau)^T \Upsilon_a(\tau,\theta) x_a^\tau + 2(x_a^\tau)^T \eta_a(\tau,\theta)\}$ and $\upsilon_a(\tau,\theta) \triangleq \ln\{\rho_a(\tau,\theta)\}$ for some $\theta \in \mathbb{R}^+$ and $\forall \tau \in [t_0, t_f]$. Then, the cumulant-generating function is given by*

$$\psi(\tau, x_a^\tau; \theta) = (x_a^\tau)^T \Upsilon_a(\tau,\theta) x_a + 2(x_a^\tau)^T \eta_a(\tau,\theta) + \upsilon_a(\tau,\theta) \tag{6.18}$$

where the solutions $\Upsilon_a(\tau,\theta)$, $\eta_a(\tau,\theta)$, and $\upsilon_a(\tau,\theta)$ solve the backward-in-time matrix-valued differential equation

$$\frac{\mathrm{d}}{\mathrm{d}\tau}\Upsilon_a(\tau,\theta) = -A_a^T(\tau)\Upsilon_a(\tau,\theta) - \Upsilon_a(\tau,\theta)A_a(\tau) - 2\Upsilon_a(\tau,\theta)G_a(\tau)W_a G_a^T(\tau)\Upsilon_a(\tau,\theta) - \theta Q_a(\tau) \tag{6.19}$$

the backward-in-time vector-valued differential equation

$$\frac{\mathrm{d}}{\mathrm{d}\tau}\eta_a(\tau,\theta) = -A_a^T(\tau)\eta_a(\tau,\theta) - \Upsilon_a(\tau,\theta)B_a(\tau)v(\tau) - \Upsilon_a(\tau,\theta)\Gamma_a(\tau)u_r(\tau) - \theta N_a(\tau)v(\tau) \tag{6.20}$$

and the backward-in-time scalar-valued differential equation

$$\frac{\mathrm{d}}{\mathrm{d}\tau}\upsilon_a(\tau,\theta) = -\operatorname{Tr}\{\Upsilon_a(\tau,\theta)G_a(\tau)W_a G_a^T(\tau)\} - 2\eta_a^T(\tau,\theta)(B_a(\tau)v(\tau) + \Gamma_a(\tau)u_r(\tau)) - \theta v^T(\tau)R(\tau)v(\tau) \tag{6.21}$$

with the terminal-value conditions $\Upsilon_a(t_f,\theta) = \theta Q_{fa}$, $\eta_a(t_f,\theta) = 0$, $\upsilon_a(t_f,\theta) = 0$.

By definition, cumulants that provide performance information for the decision process taken by the model-following control design can best be generated by the Maclaurin series expansion of the cumulant generating function (6.18):

$$\psi(\tau, x_a^\tau; \theta) \triangleq \sum_{r=1}^{\infty} \kappa_r \frac{\theta^r}{r!} = \sum_{r=1}^{\infty} \frac{\partial^r}{\partial \theta^r}\psi(\tau, x_a^\tau; \theta)\bigg|_{\theta=0} \frac{\theta^r}{r!},$$

where κ_r are denoted as rth-order cumulants or performance-measure statistics.

In view of the cumulant-generating function (6.18), performance-measure statistics that measure amount of information about performance variations in the performance measure (6.17) are obtained

$$\kappa_r = (x_a^\tau)^T \frac{\partial^r}{\partial \theta^r}\Upsilon_a(\tau,\theta)\bigg|_{\theta=0} x_a^\tau + 2(x_a^\tau)^T \frac{\partial^r}{\partial \theta^r}\eta_a(\tau,\theta)\bigg|_{\theta=0} + \frac{\partial^r}{\partial \theta^r}\upsilon_a(\tau,\theta)\bigg|_{\theta=0}. \tag{6.22}$$

For notational convenience, it is necessary to introduce

$$H_a(\tau,r) \triangleq \frac{\partial^r}{\partial\theta^r}\Upsilon_a(\tau,\theta)\bigg|_{\theta=0}, \quad D_a(\tau,r) \triangleq \frac{\partial^r}{\partial\theta^r}\eta_a(\tau,\theta)\bigg|_{\theta=0},$$

$$d_a(\tau,r) \triangleq \frac{\partial^r}{\partial\theta^r}\upsilon_a(\tau,\theta)\bigg|_{\theta=0}.$$

Therefore, the next result whose proof is omitted here, provides measures of the amount, value, and the design of performance information structures.

Theorem 6.2.2 (Performance-Measure Statistics). *In spirit of the system considered in Eqs. (6.16)–(6.17), the kth-order cumulant or performance-measure statistic in the problem of model following is given by*

$$\kappa_k = x_{a0}^T H_a(t_0,k)x_{a0} + 2x_{a0}^T D_a(t_0,k) + d_a(t_0,k), \tag{6.23}$$

whereby $\{H_a(\tau,r)\}_{r=1}^k$, $\{D_a(\tau,r)\}_{r=1}^k$, *and* $\{d_a(\tau,r)\}_{r=1}^k$ *evaluated at* $\tau = t_0$ *satisfy the backward-in-time differential equations (with the dependence of* $H_a(\tau,r)$, $D_a(\tau,r)$, *and* $d_a(\tau,r)$ *upon* K_x, K_{z_1}, *and* v *suppressed)*

$$\frac{d}{d\tau}H_a(\tau,1) = -A_a^T(\tau)H_a(\tau,1) - H_a(\tau,1)A_a(\tau) - Q_a(\tau), \tag{6.24}$$

$$\frac{d}{d\tau}H_a(\tau,r) = -A_a^T(\tau)H_a(\tau,r) - H_a(\tau,r)A_a(\tau)$$
$$- \sum_{s=1}^{r-1}\frac{2r!}{s!(r-s)!}H_a(\tau,s)G_a(\tau)W_aG_a^T(\tau)H_a(\tau,r-s), \quad r \ge 2, \tag{6.25}$$

$$\frac{d}{d\tau}D_a(\tau,1) = -A_a^T(\tau)D_a(\tau,1) - H_a(\tau,1)B_a(\tau)v(\tau)$$
$$- H_a(\tau,1)\Gamma_a(\tau)u_r(\tau) - N_a(\tau)v(\tau), \tag{6.26}$$

$$\frac{d}{d\tau}D_a(\tau,r) = -A_a^T(\tau)D_a(\tau,r) - H_a(\tau,r)B_a(\tau)v(\tau)$$
$$- H_a(\tau,r)\Gamma_a(\tau)u_r(\tau), \quad 2 \le r \le k, \tag{6.27}$$

$$\frac{d}{d\tau}d_a(\tau,1) = -\operatorname{Tr}\{H_a(\tau,1)G_a(\tau)W_aG_a^T(\tau)\}$$
$$- 2D_a^T(\tau,1)(B_a(\tau)v(\tau)+\Gamma_a(\tau)u_r(\tau)) - v^T(\tau)R(\tau)v(\tau), \tag{6.28}$$

$$\frac{d}{d\tau}d_a(\tau,r) = -\operatorname{Tr}\{H_a(\tau,r)G_a(\tau)W_aG_a^T(\tau)\}$$
$$- 2D_a^T(\tau,r)(B_a(\tau)v(\tau)+\Gamma_a(\tau)u_r(\tau)), \quad 2 \le r \le k, \tag{6.29}$$

where the terminal-value conditions $H_a(t_f,1) = Q_{fa}$, $H_a(t_f,r) = 0$ *for* $2 \le r \le k$, $D_a(t_f,i) = 0$ *for* $1 \le r \le k$ *and* $d_a(t_f,r) = 0$ *for* $1 \le r \le k$.

In the design of a decision process in which the information process about performance variations is embedded with K_x, K_{z_1}, and v, it is convenient to rewrite the results (6.24)–(6.29) in accordance of the following matrix and vector partitions

$$H_a(\tau,r) = \begin{bmatrix} \mathscr{H}_{11}^r(\tau) & \mathscr{H}_{12}^r(\tau) \\ (\mathscr{H}_{12}^r)^T(\tau) & \mathscr{H}_{22}^r(\tau) \end{bmatrix} \text{ and } D_a(\tau,r) = \begin{bmatrix} \mathscr{D}_{11}^r(\tau) \\ \mathscr{D}_{21}^r(\tau) \end{bmatrix}.$$

Corollary 6.2.1 (Alternate Representations of Performance-Measure Statistics). *Let the system be governed by Eqs. (6.16)–(6.17). Then, the kth cumulant of performance measure (6.17) is also given by*

$$\kappa_k = x_0^T \mathscr{H}_{11}^k(t_0)x_0 + 2x_0^T \mathscr{H}_{12}^k(t_0)z_{10} + z_{10}^T \mathscr{H}_{22}^k(t_0)z_{10} + \mathscr{D}^k(t_0) + 2x_0^T \mathscr{D}_{11}^k(t_0) + 2z_{10}^T \mathscr{D}_{21}^k(t_0), \quad k \in \mathbb{Z}^+, \tag{6.30}$$

where bounded solutions $\{\mathscr{H}_{11}^r(\tau)\}_{r=1}^k$, $\{\mathscr{H}_{12}^r(\tau)\}_{r=1}^k$, $\{\mathscr{H}_{22}^r(\tau)\}_{r=1}^k$, $\{\mathscr{D}_{11}^r(\tau)\}_{r=1}^k$, $\{\mathscr{D}_{21}^r(\tau)\}_{r=1}^k$, and $\{\mathscr{D}^r(\tau)\}_{r=1}^k$ satisfy the time-backward differential equations

$$\begin{aligned}\frac{d}{d\tau}\mathscr{H}_{11}^1(\tau) &\triangleq \mathscr{F}_{11}^1(\tau,\mathscr{H}_{11},K_x), \\ &= -[A(\tau)+B(\tau)K_x(\tau)]^T \mathscr{H}_{11}^1(\tau) - \mathscr{H}_{11}^1(\tau)[A(\tau)+B(\tau)K_x(\tau)] \\ &\quad - K_x^T(\tau)R(\tau)K_x(\tau) - C^T(\tau)Q(\tau)C(\tau),\end{aligned} \tag{6.31}$$

$$\begin{aligned}\frac{d}{d\tau}\mathscr{H}_{11}^r(\tau) &\triangleq \mathscr{F}_{11}^r(\tau,\mathscr{H}_{11},K_x) \\ &= -[A(\tau)+B(\tau)K_x(\tau)]^T \mathscr{H}_{11}^r(\tau) - \mathscr{H}_{11}^r(\tau)[A(\tau)+B(\tau)K_x(\tau)] \\ &\quad - \sum_{s=1}^{r-1} \frac{2r!}{s!(r-s)!} \mathscr{H}_{11}^s(\tau)L(\tau)WL^T(\tau)\mathscr{H}_{11}^{r-s}(\tau) \\ &\quad - \sum_{s=1}^{r-1} \frac{2r!}{s!(r-s)!} \mathscr{H}_{12}^s(\tau)L_1(\tau)W_1L_1^T(\tau)(\mathscr{H}_{12}^{r-s})^T(\tau), \quad 2 \le r \le k,\end{aligned} \tag{6.32}$$

$$\begin{aligned}\frac{d}{d\tau}\mathscr{H}_{12}^1(\tau) &\triangleq \mathscr{F}_{12}^1(\tau,\mathscr{H}_{12},\mathscr{H}_{11},K_x,K_{z_1}) \\ &= -[A(\tau)+B(\tau)K_x(\tau)]^T \mathscr{H}_{12}^1(\tau) - K_x^T(\tau)R(\tau)K_{z_1}(\tau) \\ &\quad - \mathscr{H}_{11}^1(\tau)B(\tau)K_{z_1}(\tau) - \mathscr{H}_{12}^1(\tau)A_1(\tau) + C^T(\tau)Q(\tau)C_1(\tau),\end{aligned} \tag{6.33}$$

$$\begin{aligned}\frac{d}{d\tau}\mathscr{H}_{12}^r(\tau) &\triangleq \mathscr{F}_{12}^r(\tau,\mathscr{H}_{12},\mathscr{H}_{11},K_x,K_{z_1}) \\ &= -[A(\tau)+B(\tau)K_x(\tau)]^T \mathscr{H}_{12}^r(\tau) - \mathscr{H}_{11}^r(\tau)B(\tau)K_{z_1}(\tau) - \mathscr{H}_{12}^r(\tau)A_1(\tau)\end{aligned}$$

$$-\sum_{s=1}^{r-1}\frac{2r!}{s!(r-s)!}\mathscr{H}_{11}^{s}(\tau)L(\tau)WL^{T}(\tau)\mathscr{H}_{12}^{r-s}(\tau)$$
$$-\sum_{s=1}^{r-1}\frac{2r!}{s!(r-s)!}\mathscr{H}_{12}^{s}(\tau)L_{1}(\tau)W_{1}L_{1}^{T}(\tau)\mathscr{H}_{22}^{r-s}(\tau), \tag{6.34}$$

$$\frac{\mathrm{d}}{\mathrm{d}\tau}\mathscr{H}_{22}^{1}(\tau) \triangleq \mathscr{F}_{22}^{1}(\tau,\mathscr{H}_{22},\mathscr{H}_{12},K_{z_1})$$
$$= -A_1^T(\tau)\mathscr{H}_{22}^{1}(\tau)-\mathscr{H}_{22}^{1}(\tau)A_1(\tau)-K_{z_1}^T(\tau)B^T(\tau)\mathscr{H}_{12}^{1}(\tau)$$
$$-(\mathscr{H}_{12}^{1})^T(\tau)B(\tau)K_{z_1}(\tau)-K_{z_1}^T(\tau)R(\tau)K_{z_1}(\tau)-C_1^T(\tau)Q(\tau)C_1(\tau), \tag{6.35}$$

$$\frac{\mathrm{d}}{\mathrm{d}\tau}\mathscr{H}_{22}^{r}(\tau) \triangleq \mathscr{F}_{22}^{r}(\tau,\mathscr{H}_{22},\mathscr{H}_{12},K_{z_1})$$
$$= -A_1^T(\tau)\mathscr{H}_{22}^{r}(\tau)-\mathscr{H}_{22}^{r}(\tau)A_1(\tau)-K_{z_1}^T(\tau)B^T(\tau)\mathscr{H}_{12}^{r}(\tau)$$
$$-(\mathscr{H}_{12}^{r})^T(\tau)B(\tau)K_{z_1}(\tau)$$
$$-\sum_{s=1}^{r-1}\frac{2r!}{s!(r-s)!}(\mathscr{H}_{12}^{s})^T(\tau)L(\tau)WL^{T}(\tau)\mathscr{H}_{12}^{r-s}(\tau)$$
$$-\sum_{s=1}^{r-1}\frac{2r!}{s!(r-s)!}\mathscr{H}_{22}^{s}(\tau)L_{1}(\tau)W_{1}L_{1}^{T}(\tau)\mathscr{H}_{22}^{r-s}(\tau),\quad 2\le r\le k, \tag{6.36}$$

$$\frac{\mathrm{d}}{\mathrm{d}\tau}\mathscr{D}_{11}^{1}(\tau) \triangleq \mathscr{G}_{11}^{1}(\tau,\mathscr{D}_{11},\mathscr{H}_{11},\mathscr{H}_{12},K_x,v)$$
$$= -[A(\tau)+B(\tau)K_x(\tau)]^T\,\mathscr{D}_{11}^{1}(\tau)-\mathscr{H}_{11}^{1}(\tau)B(\tau)v(\tau)$$
$$-\mathscr{H}_{12}^{1}(\tau)B_1(\tau)u_r(\tau)-K_x^T(\tau)R(\tau)v(\tau), \tag{6.37}$$

$$\frac{\mathrm{d}}{\mathrm{d}\tau}\mathscr{D}_{11}^{r}(\tau) \triangleq \mathscr{G}_{11}^{r}(\tau,\mathscr{D}_{11},\mathscr{H}_{11},\mathscr{H}_{12},K_x,v)$$
$$= -[A(\tau)+B(\tau)K_x(\tau)]^T\,\mathscr{D}_{11}^{r}(\tau)-\mathscr{H}_{11}^{r}(\tau)B(\tau)v(\tau)$$
$$-\mathscr{H}_{12}^{r}(\tau)B_1(\tau)u_r(\tau),\quad 2\le r\le k, \tag{6.38}$$

$$\frac{\mathrm{d}}{\mathrm{d}\tau}\mathscr{D}_{21}^{1}(\tau) \triangleq \mathscr{G}_{21}^{1}(\tau,\mathscr{D}_{21},\mathscr{D}_{11},\mathscr{H}_{12},\mathscr{H}_{22},K_{z_1},v)$$
$$= -K_{z_1}^T(\tau)B^T(\tau)\mathscr{D}_{11}^{1}(\tau)-(\mathscr{H}_{12}^{1})^T(\tau)B(\tau)v(\tau)$$
$$-A_1^T(\tau)\mathscr{D}_{21}^{1}(\tau)-\mathscr{H}_{22}^{1}(\tau)B_1(\tau)u_r(\tau)-K_{z_1}^T(\tau)R(\tau)v(\tau), \tag{6.39}$$

$$\frac{\mathrm{d}}{\mathrm{d}\tau}\mathscr{D}_{21}^{r}(\tau) \triangleq \mathscr{G}_{21}^{1}(\tau,\mathscr{D}_{21},\mathscr{D}_{11},\mathscr{H}_{12},\mathscr{H}_{22},K_{z_1},v)$$
$$= -K_{z_1}^T(\tau)B^T(\tau)\mathscr{D}_{11}^{r}(\tau)-(\mathscr{H}_{12}^{r})^T(\tau)B(\tau)v(\tau)$$
$$-A_1^T(\tau)\mathscr{D}_{21}^{r}(\tau)-\mathscr{H}_{22}^{r}(\tau)B_1(\tau)u_r(\tau),\quad 2\le r\le k, \tag{6.40}$$

$$\begin{aligned}\frac{d}{d\tau}\mathscr{D}^1(\tau) &\triangleq \mathscr{G}^1(\tau,\mathscr{H}_{11},\mathscr{D}_{11},\mathscr{D}_{21},v)\\ &= -\mathrm{Tr}\{\mathscr{H}_{11}^1(\tau)L(\tau)WL^T(\tau)\} - \mathrm{Tr}\{\mathscr{H}_{22}^1(\tau)L_1(\tau)W_1L_1^T(\tau)\}\\ &\quad -2(\mathscr{D}_{11}^1)^T(\tau)B(\tau)v(\tau) - 2(\mathscr{D}_{21}^1)^T(\tau)B_1(\tau)u_r(\tau) - v^T(\tau)R(\tau)v(\tau),\end{aligned} \tag{6.41}$$

$$\begin{aligned}\frac{d}{d\tau}\mathscr{D}^r(\tau) &\triangleq \mathscr{G}^r(\tau,\mathscr{H}_{11},\mathscr{D}_{11},\mathscr{D}_{21},v)\\ &= -\mathrm{Tr}\{\mathscr{H}_{11}^r(\tau)L(\tau)WL^T(\tau)\} - \mathrm{Tr}\{\mathscr{H}_{22}^r(\tau)L_1(\tau)W_1L_1^T(\tau)\}\\ &\quad -2(\mathscr{D}_{11}^r)^T(\tau)B(\tau)v(\tau) - 2(\mathscr{D}_{21}^r)^T(\tau)B_1(\tau)u_r(\tau), \quad 2\le r\le k.\end{aligned} \tag{6.42}$$

Moreover, the terminal-value conditions are $\mathscr{H}_{11}^1(t_f)=C^T(t_f)Q_fC(t_f)$, $\mathscr{H}_{11}^r(t_f)=0$ *for* $2\le r\le k$; $\mathscr{H}_{12}^1(t_f)=-C^T(t_f)Q_fC_1(t_f)$, $\mathscr{H}_{12}^r(t_f)=0$ *for* $2\le r\le k$; $\mathscr{H}_{22}^1(t_f)=C_1^T(t_f)Q_fC_1(t_f)$, $\mathscr{H}_{22}^r(t_f)=0$ *for* $2\le r\le k$; $\mathscr{D}_{11}^r(t_f)=0$ *for* $1\le r\le k$; $\mathscr{D}_{21}^r(t_f)=0$ *for* $1\le r\le k$; $\mathscr{D}^r(t_f)=0$ *for* $1\le r\le k$. *And the components of the k-tuple variables* $\mathscr{H}_{11}$, $\mathscr{H}_{12}$, $\mathscr{H}_{22}$, $\mathscr{D}_{11}$, $\mathscr{D}_{21}$, *and* $\mathscr{D}$ *are defined by*

$$\mathscr{H}_{11}\triangleq(\mathscr{H}_{11}^1,\ldots,\mathscr{H}_{11}^k),\quad \mathscr{H}_{12}\triangleq(\mathscr{H}_{12}^1,\ldots,\mathscr{H}_{12}^k),\quad \mathscr{H}_{22}\triangleq(\mathscr{H}_{22}^1,\ldots,\mathscr{H}_{22}^k),$$
$$\mathscr{D}_{11}\triangleq(\mathscr{D}_{11}^1,\ldots,\mathscr{D}_{11}^k),\quad \mathscr{D}_{21}\triangleq(\mathscr{D}_{21}^1,\ldots,\mathscr{D}_{21}^k),\quad \mathscr{D}\triangleq(\mathscr{D}^1,\ldots,\mathscr{D}^k).$$

For a rigorous basis of a well-posed optimization problem, the following sufficient conditions for the existence of solutions to Eqs. (6.31)–(6.42) are stated.

Corollary 6.2.2 (Existence of Performance-Measure Statistics). *In addition to asymptotically stable reference models, it is assumed that* (A,B) *and* (A_1,B_1) *are uniformly stabilizable;* (A,C) *and* (A_1,C_1) *are uniformly detectable. Then, for any given* $k\in\mathbb{N}$, *the backward-in-time matrix differential equations (6.31)–(6.42) admit unique and bounded solutions* $\{\mathscr{H}_{11}(\tau)\}_{i=1}^k$, $\{\mathscr{H}_{12}(\tau)\}_{i=1}^k$, $\{\mathscr{H}_{22}(\tau)\}_{i=1}^k$, $\{\mathscr{D}_{11}(\tau)\}_{i=1}^k$, $\{\mathscr{D}_{21}(\tau)\}_{i=1}^k$ *and* $\{\mathscr{D}(\tau)\}_{i=1}^k$ *on* $[t_0,t_f]$.

6.3 Performance-Measure Statistics for Risk-Averse Control

To formulate in precise terms for optimization problem in the model-following control design, it is important to note that all cumulants (6.30) are the functions of backward-in-time evolutions governed by Eqs. (6.31)–(6.42) and initial valued states $x(t_0)$ and $z_1(t_0)$. Henceforth, these backward-in-time trajectories (6.31)–(6.42) are therefore considered as the new dynamical equations together with state variables $\{\mathscr{H}_{11}^r(\cdot)\}_{i=1}^k$, $\{\mathscr{H}_{12}^r(\cdot)\}_{i=1}^k$, $\{\mathscr{H}_{22}^r(\cdot)\}_{i=1}^k$, $\{\mathscr{D}_{11}^r(\cdot)\}_{i=1}^k$, $\{\mathscr{D}_{21}^r(\cdot)\}_{i=1}^k$ and $\{\mathscr{D}^r(\cdot)\}_{i=1}^k$; and thus, not the traditional system state estimates $\hat{x}(\cdot)$ and $\hat{z}_1(\cdot)$.

For a compact formulation, the dynamical equations (6.31)–(6.42) can be rewritten by the product mappings

$$\frac{d}{d\tau}\mathscr{H}_{11}(\tau) = \mathscr{F}_{11}(\tau, \mathscr{H}_{11}, K_x), \quad \mathscr{H}_{11}(t_f) = \mathscr{H}_{11}^f, \tag{6.43}$$

$$\frac{d}{d\tau}\mathscr{H}_{12}(\tau) = \mathscr{F}_{12}(\tau, \mathscr{H}_{12}, \mathscr{H}_{11}, K_x, K_{z_1}), \quad \mathscr{H}_{12}(t_f) = \mathscr{H}_{12}^f, \tag{6.44}$$

$$\frac{d}{d\tau}\mathscr{H}_{22}(\tau) = \mathscr{F}_{22}(\tau, \mathscr{H}_{22}, \mathscr{H}_{12}, K_{z_1}), \quad \mathscr{H}_{22}(t_f) = \mathscr{H}_{22}^f, \tag{6.45}$$

$$\frac{d}{d\tau}\mathscr{D}_{11}(\tau) = \mathscr{G}_{11}(\tau, \mathscr{D}_{11}, \mathscr{H}_{11}, \mathscr{H}_{12}, K_x, v), \quad \mathscr{D}_{11}(t_f) = \mathscr{D}_{11}^f, \tag{6.46}$$

$$\frac{d}{d\tau}\mathscr{D}_{21}(\tau) = \mathscr{G}_{21}(\tau, \mathscr{D}_{21}, \mathscr{D}_{11}, \mathscr{H}_{12}, \mathscr{H}_{22}, K_{z_1}, v), \quad \mathscr{D}_{21}(t_f) = \mathscr{D}_{21}^f, \tag{6.47}$$

$$\frac{d}{d\tau}\mathscr{D}(\tau) = \mathscr{G}(\tau, \mathscr{H}_{11}, \mathscr{D}_{11}, \mathscr{D}_{21}, v), \quad \mathscr{D}(t_f) = \mathscr{D}^f \tag{6.48}$$

under the obvious definitions of Cartesian product mappings

$$\mathscr{F}_{11} \triangleq \mathscr{F}_{11}^1 \times \cdots \times \mathscr{F}_{11}^k, \quad \mathscr{F}_{12} \triangleq \mathscr{F}_{12}^1 \times \cdots \times \mathscr{F}_{12}^k, \quad \mathscr{F}_{22} \triangleq \mathscr{F}_{22}^1 \times \cdots \times \mathscr{F}_{22}^k,$$
$$\mathscr{G}_{11} \triangleq \mathscr{G}_{11}^1 \times \cdots \times \mathscr{G}_{11}^k, \quad \mathscr{G}_{21} \triangleq \mathscr{G}_{21}^1 \times \cdots \times \mathscr{G}_{21}^k, \quad \mathscr{G} \triangleq \mathscr{G}^1 \times \cdots \times \mathscr{G}^k$$

and the terminal values of the resultant state variables $\mathscr{H}_{11}(\cdot)$, $\mathscr{H}_{12}(\cdot)$, $\mathscr{H}_{22}(\cdot)$, $\mathscr{D}_{11}(\cdot)$, $\mathscr{D}_{21}(\cdot)$, and $\mathscr{D}(\cdot)$ are given by

$$\mathscr{H}_{11}^f \triangleq (C^T(t_f)Q_fC(t_f), 0, \ldots, 0), \quad \mathscr{D}_{11}^f \triangleq (0, \ldots, 0),$$
$$\mathscr{H}_{12}^f \triangleq (-C^T(t_f)Q_fC_1(t_f), 0, \ldots, 0), \quad \mathscr{D}_{21}^f \triangleq (0, \ldots, 0),$$
$$\mathscr{H}_{22}^f \triangleq (C_1^T(t_f)Q_fC_1(t_f), 0, \ldots, 0), \quad \mathscr{D}^f \triangleq (0, \ldots, 0).$$

As before, the product system (6.43)–(6.48) uniquely determines the state variables $\mathscr{H}_{11}(\cdot)$, $\mathscr{H}_{12}(\cdot)$, $\mathscr{H}_{22}(\cdot)$, $\mathscr{D}_{11}(\cdot)$, $\mathscr{D}_{21}(\cdot)$ and $\mathscr{D}(\cdot)$ once the admissible feedback gain $K_x(\cdot)$, feedforward gain $K_{z_1}(\cdot)$, and affine input $v(\cdot)$ are specified. Therefore, the state variables $\mathscr{H}_{11}(\cdot)$, $\mathscr{H}_{12}(\cdot)$, $\mathscr{H}_{22}(\cdot)$, $\mathscr{D}_{11}(\cdot)$, $\mathscr{D}_{21}(\cdot)$ and $\mathscr{D}(\cdot)$ are considered as $\mathscr{H}_{11}(\cdot, K_x, K_{z_1}, v)$, $\mathscr{H}_{12}(\cdot, K_x, K_{z_1}, v)$, $\mathscr{H}_{22}(\cdot, K_x, K_{z_1}, v)$, $\mathscr{D}_{11}(\cdot, K_x, K_{z_1}, v)$, $\mathscr{D}_{21}(\cdot, K_x, K_{z_1}, v)$ and $\mathscr{D}(\cdot, K_x, K_{z_1}, v)$, respectively.

In the light of performance risks and stochastic preferences, the generalized performance index that follows for the reliable control of model-following stochastic systems provides some tradeoffs on what means for performance riskiness from the standpoint of higher-order characteristics pertaining to performance sampling distributions.

Definition 6.3.1 (Risk-Value Aware Performance Index). Fix $k \in \mathbb{Z}^+$ and the sequence $\mu = \{\mu_i \geq 0\}_{i=1}^k$ with $\mu_1 > 0$. Then, the performance index of Mayer type with value and risk considerations is given by

$$\begin{aligned}
&\phi_m\left(t_0, \mathscr{H}_{11}(t_0), \mathscr{H}_{12}(t_0), \mathscr{H}_{22}(t_0), \mathscr{D}_{11}(t_0), \mathscr{D}_{21}(t_0), \mathscr{D}(t_0)\right) \\
&\triangleq \underbrace{\mu_1 \kappa_1}_{\text{Value Measure}} + \underbrace{\mu_2 \kappa_2 + \cdots + \mu_k \kappa_k}_{\text{Risk Measures}} \\
&= \sum_{r=1}^{k} \mu_r \Big[x_0^T \mathscr{H}_{11}^r(t_0) x_0 + 2x_0^T \mathscr{H}_{12}^r(t_0) z_{10} + z_{10}^T \mathscr{H}_{22}^r(t_0) z_{10} \\
&\quad + 2x_0^T \mathscr{D}_{11}^r(t_0) + 2z_{10}^T \mathscr{D}_{21}^r(t_0) + \mathscr{D}^r(t_0) \Big], \qquad (6.49)
\end{aligned}$$

where the parametric design of freedom μ_r chosen by the control designer represents preferential weights on simultaneous higher-order characteristics, e.g., mean (i.e., the average of performance measure), variance (i.e., the dispersion of values of performance measure around its mean), and skewness (i.e., the anti-symmetry of the probability density of performance measure), pertaining to closed-loop performance variations and uncertainties, whereas the k-tuple cumulant variables $\mathscr{H}_{11}$, $\mathscr{H}_{12}$, $\mathscr{H}_{22}$, $\mathscr{D}_{11}$, $\mathscr{D}_{21}$, and $\mathscr{D}$ evaluated at $\tau = t_0$ satisfy the new dynamical equations (6.43)–(6.48).

For terminal data $(t_f, \mathscr{H}_{11}^f, \mathscr{H}_{12}^f, \mathscr{H}_{22}^f, \mathscr{D}_{11}^f, \mathscr{D}_{21}^f, \mathscr{D}^f)$ given, the classes of admissible feedback, feedforward gains and affine inputs are then defined as follows.

Definition 6.3.2 (Admissible Gains and Affine Inputs). Let compact subsets $\overline{J}_x \subset \mathbb{R}^{m\times n}$, $\overline{J}_{z_1} \subset \mathbb{R}^{m\times n_1}$, and $\overline{V} \subset \mathbb{R}^m$ be the allowable sets of gains and affine input values. For the given $k \in \mathbb{N}$ and the sequence $\mu = \{\mu_r \geq 0\}_{r=1}^k$ with $\mu_1 > 0$, $\mathscr{K}^x_{t_f, \mathscr{H}_{11}^f, \mathscr{H}_{12}^f, \mathscr{D}_{11}^f, \mathscr{D}_{21}^f, \mathscr{H}_{22}^f, \mathscr{D}^f; \mu}$, $\mathscr{K}^{z_1}_{t_f, \mathscr{H}_{11}^f, \mathscr{H}_{12}^f, \mathscr{H}_{22}^f, \mathscr{D}_{11}^f, \mathscr{D}_{21}^f, \mathscr{D}^f; \mu}$ and $\mathscr{V}_{t_f, \mathscr{H}_{11}^f, \mathscr{H}_{12}^f, \mathscr{H}_{22}^f, \mathscr{D}_{11}^f, \mathscr{D}_{21}^f, \mathscr{D}^f; \mu}$ are the classes of $\mathscr{C}([t_0, t_f]; \mathbb{R}^{m\times n})$, $\mathscr{C}([t_0, t_f]; \mathbb{R}^{m\times n_1})$, and $\mathscr{C}([t_0, t_f]; \mathbb{R}^m)$ with values $K_x(\cdot) \in \overline{K}_x$, $K_{z_1}(\cdot) \in \overline{K}_{z_1}$, and $v(\cdot) \in \overline{V}$ for which unique and bounded solutions to the equations (6.43)–(6.48) exist on $[t_0, t_f]$.

Next, the optimization statements for risk-averse control of the model-following problem over a finite horizon are stated.

Definition 6.3.3 (Optimization for Mayer Problem). Suppose $k \in \mathbb{N}$ and the sequence $\mu = \{\mu_i \geq 0\}_{i=1}^k$ with $\mu_1 > 0$ are fixed. Then, the optimization problem for risk-averse control of model-following systems is defined by the minimization of value and risk-aware performance index (6.49) over the admissible sets $\mathscr{K}^x_{t_f, \mathscr{H}_{11}^f, \mathscr{H}_{12}^f, \mathscr{H}_{22}^f, \mathscr{D}_{11}^f, \mathscr{D}_{21}^f, \mathscr{D}^f; \mu}$, $\mathscr{K}^{z_1}_{t_f, \mathscr{H}_{11}^f, \mathscr{H}_{12}^f, \mathscr{H}_{22}^f, \mathscr{D}_{11}^f, \mathscr{D}_{21}^f, \mathscr{D}^f; \mu}$, $\mathscr{V}_{t_f, \mathscr{H}_{11}^f, \mathscr{H}_{12}^f, \mathscr{H}_{22}^f, \mathscr{D}_{11}^f, \mathscr{D}_{21}^f, \mathscr{D}^f; \mu}$ and subject to the dynamics (6.43)–(6.48) on $[t_0, t_f]$.

Upon developing a value function which is defined as the value of the performance index starting at $(\varepsilon, \mathscr{Y}_{11}, \mathscr{Y}_{12}, \mathscr{Y}_{22}, \mathscr{Z}_{11}, \mathscr{Z}_{21}, \mathscr{Z})$ and proceeding optimally to the specified terminal conditions, the concept of a reachable set to which the terminal state and time, i.e., $(\varepsilon, \mathscr{Y}_{11}, \mathscr{Y}_{12}, \mathscr{Y}_{22}, \mathscr{Z}_{11}, \mathscr{Z}_{21}, \mathscr{Z})$, resides, is necessary.

Definition 6.3.4 (Reachable Set). Let $\mathscr{Q} \triangleq \Big\{(\varepsilon, \mathscr{Y}_{11}, \mathscr{Y}_{12}, \mathscr{Y}_{22}, \mathscr{Z}_{11}, \mathscr{Z}_{21}, \mathscr{Z}) \in [t_0,t_f] \times (\mathbb{R}^n)^k \times (\mathbb{R}^{n\times n_1})^k \times (\mathbb{R}^{n_1})^k \times (\mathbb{R}^n)^k \times (\mathbb{R}^{n_1})^k \times \mathbb{R}^k$ so $\mathscr{K}^x_{t_f,\mathscr{H}^f_{11},\mathscr{H}^f_{12},\mathscr{H}^f_{22},\mathscr{D}^f_{11},\mathscr{D}^f_{21},\mathscr{D}^f;\mu} \times \mathscr{K}^{z_1}_{t_f,\mathscr{H}^f_{11},\mathscr{H}^f_{12},\mathscr{H}^f_{22},\mathscr{D}^f_{11},\mathscr{D}^f_{21},\mathscr{D}^f;\mu} \times \mathscr{V}_{t_f,\mathscr{H}^f_{11},\mathscr{H}^f_{12},\mathscr{H}^f_{22},\mathscr{D}^f_{11},\mathscr{D}^f_{21},\mathscr{D}^f;\mu} \neq \emptyset \Big\}$.

The HJB equation satisfied by the value function $\mathscr{V}(\varepsilon, \mathscr{Y}_{11}, \mathscr{Y}_{12}, \mathscr{Y}_{22}, \mathscr{Z}_{11}, \mathscr{Z}_{21}, \mathscr{Z})$ as the infimum of the performance index (6.49) over $\mathscr{K}^x_{\varepsilon,\mathscr{Y}_{11},\mathscr{Y}_{12},\mathscr{Y}_{22},\mathscr{Z}_{11},\mathscr{Z}_{21},\mathscr{Z};\mu}$, $\mathscr{K}^{z_1}_{\varepsilon,\mathscr{Y}_{11},\mathscr{Y}^f_{12},\mathscr{Y}_{22},\mathscr{Z}^f_{11},\mathscr{Z}^f_{21},\mathscr{Z}^f;\mu}$ and $\mathscr{V}_{\varepsilon,\mathscr{Y}_{11},\mathscr{Y}_{12},\mathscr{Y}_{22},\mathscr{Z}_{11},\mathscr{Z}_{21},\mathscr{Z};\mu}$ is now described as below.

Theorem 6.3.1 (HJB Equation for Mayer Problem). *Let $(\varepsilon, \mathscr{Y}_{11}, \mathscr{Y}_{12}, \mathscr{Y}_{22}, \mathscr{Z}_{11}, \mathscr{Z}_{21}, \mathscr{Z})$ be any interior point of the reachable set $\mathscr{Q}$ at which the value function $\mathscr{V}(\varepsilon, \mathscr{Y}_{11}, \mathscr{Y}_{12}, \mathscr{Y}_{22}, \mathscr{Z}_{11}, \mathscr{Z}_{21}, \mathscr{Z})$ is differentiable. If there exist optimal $K_x^* \in \mathscr{K}^x_{t_f,\mathscr{H}^f_{11},\mathscr{H}^f_{12},\mathscr{H}^f_{22},\mathscr{D}^f_{11},\mathscr{D}^f_{21},\mathscr{D}^f;\mu}$, $K_{z_1}^* \in \mathscr{K}^{z_1}_{t_f,\mathscr{H}^f_{11},\mathscr{H}^f_{12},\mathscr{H}^f_{22},\mathscr{D}^f_{11},\mathscr{D}^f_{21},\mathscr{D}^f;\mu}$ and $v \in \mathscr{V}_{t_f,\mathscr{H}^f_{11},\mathscr{H}^f_{12},\mathscr{H}^f_{22},\mathscr{D}^f_{11},\mathscr{D}^f_{21},\mathscr{D}^f;\mu}$, then the partial differential equation of dynamic programming*

$$
\begin{aligned}
0 = \min_{J_x \in \overline{K}_x, K_{z_1} \in \overline{K}_{z_1}, v \in \overline{V}} \Big\{ & \frac{\partial}{\partial \varepsilon} \mathscr{V}(\varepsilon, \mathscr{Y}_{11}, \mathscr{Y}_{12}, \mathscr{Y}_{22}, \mathscr{Z}_{11}, \mathscr{Z}_{21}, \mathscr{Z}) \\
&+ \frac{\partial}{\partial \operatorname{vec}(\mathscr{Y}_{11})} \mathscr{V}(\varepsilon, \mathscr{Y}_{11}, \mathscr{Y}_{12}, \mathscr{Y}_{22}, \mathscr{Z}_{11}, \mathscr{Z}_{21}, \mathscr{Z}) \operatorname{vec}(\mathscr{F}_{11}(\varepsilon, \mathscr{Y}_{11}, K_x)) \\
&+ \frac{\partial}{\partial \operatorname{vec}(\mathscr{Y}_{12})} \mathscr{V}(\varepsilon, \mathscr{Y}_{11}, \mathscr{Y}_{12}, \mathscr{Y}_{22}, \mathscr{Z}_{11}, \mathscr{Z}_{21}, \mathscr{Z}) \operatorname{vec}(\mathscr{F}_{12}(\varepsilon, \mathscr{Y}_{12}, \mathscr{Y}_{11}, K_x, K_{z_1})) \\
&+ \frac{\partial}{\partial \operatorname{vec}(\mathscr{Y}_{22})} \mathscr{V}(\varepsilon, \mathscr{Y}_{11}, \mathscr{Y}_{12}, \mathscr{Y}_{22}, \mathscr{Z}_{11}, \mathscr{Z}_{21}, \mathscr{Z}) \operatorname{vec}(\mathscr{F}_{22}(\varepsilon, \mathscr{Y}_{22}, \mathscr{Y}_{12}, K_{z_1})) \\
&+ \frac{\partial}{\partial \operatorname{vec}(\mathscr{Z}_{11})} \mathscr{V}(\varepsilon, \mathscr{Y}_{11}, \mathscr{Y}_{12}, \mathscr{Y}_{22}, \mathscr{Z}_{11}, \mathscr{Z}_{21}, \mathscr{Z}) \operatorname{vec}(\mathscr{G}_{11}(\varepsilon, \mathscr{Z}_{11}, \mathscr{Y}_{11}, \mathscr{Y}_{12}, K_x, v)) \\
&+ \frac{\partial}{\partial \operatorname{vec}(\mathscr{Z}_{21})} \mathscr{V}(\varepsilon, \mathscr{Y}_{11}, \mathscr{Y}_{12}, \mathscr{Y}_{22}, \mathscr{Z}_{11}, \mathscr{Z}_{21}, \mathscr{Z}) \operatorname{vec}(\mathscr{G}_{21}(\varepsilon, \mathscr{Z}_{21}, \mathscr{Z}_{11}, \mathscr{Y}_{12}, \mathscr{Y}_{22}, K_{z_1}, v)) \\
&+ \frac{\partial}{\partial \operatorname{vec}(\mathscr{Z})} \mathscr{V}(\varepsilon, \mathscr{Y}_{11}, \mathscr{Y}_{12}, \mathscr{Y}_{22}, \mathscr{Z}_{11}, \mathscr{Z}_{21}, \mathscr{Z}) \operatorname{vec}(\mathscr{G}(\varepsilon, \mathscr{Y}_{11}, \mathscr{Z}_{11}, \mathscr{Z}_{21}, v)) \Big\}
\end{aligned}
\tag{6.50}
$$

is satisfied. The boundary condition is

$$
\begin{aligned}
&\mathscr{V}(t_0, \mathscr{Y}_{11}(t_0), \mathscr{Y}_{12}(t_0), \mathscr{Y}_{22}(t_0), \mathscr{Z}_{11}(t_0), \mathscr{Z}_{21}(t_0), \mathscr{Z}(t_0)) \\
&\quad = \phi_m(t_0, \mathscr{H}_{11}(t_0), \mathscr{H}_{12}(t_0), \mathscr{H}_{22}(t_0), \mathscr{D}_{11}(t_0), \mathscr{D}_{21}(t_0), \mathscr{D}(t_0)).
\end{aligned}
$$

Note that the reachable set $\mathscr{Q}$ contains a set of points $(\varepsilon, \mathscr{Y}_{11}, \mathscr{Y}_{12}, \mathscr{Y}_{22}, \mathscr{Z}_{11}, \mathscr{Z}_{21}, \mathscr{Z})$ from which it is possible to reach the target set $\mathscr{M}$ that is a closed subset of $[t_0,t_f] \times (\mathbb{R}^n)^k \times (\mathbb{R}^{n\times n_1})^k \times (\mathbb{R}^{n_1})^k \times (\mathbb{R}^n)^k \times (\mathbb{R}^{n_1})^k \times \mathbb{R}^k$ with some trajectories corresponding to admissible feedback gain K_x, feedforward gain K_{z_1}, and affine input v. In view of $\mathscr{Q}$, there exist some candidate functions for the value function which can be actually constructed as being illustrated below.

Theorem 6.3.2 (Verification Theorem). *Fix $k \in \mathbb{N}$ and let $\mathscr{W}(\varepsilon, \mathscr{Y}_{11}, \mathscr{Y}_{12}, \mathscr{Y}_{22}, \mathscr{Z}_{11}, \mathscr{Z}_{21}, \mathscr{Z})$ be a continuously differentiable solution of the HJB equation (6.50) which satisfies the boundary condition*

$$\mathscr{W}(\varepsilon, \mathscr{Y}_{11}, \mathscr{Y}_{12}, \mathscr{Y}_{22}, \mathscr{Z}_{11}, \mathscr{Z}_{21}, \mathscr{Z}) = \phi_m(\varepsilon, \mathscr{Y}_{11}, \mathscr{Y}_{12}, \mathscr{Y}_{22}, \mathscr{Z}_{11}, \mathscr{Z}_{21}, \mathscr{Z}),$$

whereby $(\varepsilon, \mathscr{Y}_{11}, \mathscr{Y}_{12}, \mathscr{Y}_{22}, \mathscr{Z}_{11}, \mathscr{Z}_{21}, \mathscr{Z}) \in \mathscr{M}$.

Let $(t_f, \mathscr{H}_{11}^f, \mathscr{H}_{12}^f, \mathscr{H}_{22}^f, \mathscr{D}_{11}^f, \mathscr{D}_{21}^f, \mathscr{D}^f)$ be in the reachable set $\mathscr{Q}$; (K_x, K_{z_1}, v) be in $\mathscr{K}^x_{t_f, \mathscr{H}_{11}^f, \mathscr{H}_{12}^f, \mathscr{H}_{22}^f, \mathscr{D}_{11}^f, \mathscr{D}_{21}^f, \mathscr{D}^f; \mu} \times \mathscr{K}^{z_1}_{t_f, \mathscr{H}_{11}^f, \mathscr{H}_{12}^f, \mathscr{H}_{22}^f, \mathscr{D}_{11}^f, \mathscr{D}_{21}^f, \mathscr{D}^f; \mu} \times \mathscr{V}_{t_f, \mathscr{H}_{11}^f, \mathscr{H}_{12}^f, \mathscr{H}_{22}^f, \mathscr{D}_{11}^f, \mathscr{D}_{21}^f, \mathscr{D}^f; \mu}$. $\mathscr{H}_{11}$, $\mathscr{H}_{12}$, $\mathscr{H}_{22}$, $\mathscr{D}_{11}$. The corresponding solutions $\mathscr{D}_{21}$ and $\mathscr{D}$ of the dynamical equations (6.43)–(6.48). Then,. $\mathscr{W}(\tau, \mathscr{H}_{11}(\tau), \mathscr{H}_{12}(\tau), \mathscr{H}_{22}(\tau), \mathscr{D}_{11}(\tau), \mathscr{D}_{21}(\tau), \mathscr{D}(\tau))$ is a non-increasing function of τ.

If the 3-tuple $(K_x^*, K_{z_1}^*, v^*)$ is in admissible classes $\mathscr{K}^x_{t_f, \mathscr{H}_{11}^f, \mathscr{H}_{12}^f, \mathscr{H}_{22}^f, \mathscr{D}_{11}^f, \mathscr{D}_{21}^f, \mathscr{D}^f; \mu} \times \mathscr{K}^{z_1}_{t_f, \mathscr{H}_{11}^f, \mathscr{H}_{12}^f, \mathscr{H}_{22}^f, \mathscr{D}_{11}^f, \mathscr{D}_{21}^f, \mathscr{D}^f; \mu} \times \mathscr{V}_{t_f, \mathscr{H}_{11}^f, \mathscr{H}_{12}^f, \mathscr{H}_{22}^f, \mathscr{D}_{11}^f, \mathscr{D}_{21}^f, \mathscr{D}^f; \mu}$ defined on $[t_0, t_f]$ with the corresponding solutions, $\mathscr{H}_{11}^*$, $\mathscr{H}_{12}^*$, $\mathscr{H}_{22}^*$, $\mathscr{D}_{11}^*$, $\mathscr{D}_{21}^*$ and $\mathscr{D}^*$ of the dynamical equations (6.43)–(6.48) such that

$$\begin{aligned}
0 = {} & \frac{\partial}{\partial \tau} \mathscr{W}(\tau, \mathscr{H}_{11}^*(\tau), \mathscr{H}_{12}^*(\tau), \mathscr{H}_{22}^*(\tau), \mathscr{D}_{11}^*(\tau), \mathscr{D}_{21}^*(\tau), \mathscr{D}^*(\tau)) \\
& + \frac{\partial}{\partial \mathrm{vec}(\mathscr{Y}_{11})} \mathscr{W}(\tau, \mathscr{H}_{11}^*(\tau), \mathscr{H}_{12}^*(\tau), \mathscr{H}_{22}^*(\tau), \mathscr{D}_{11}^*(\tau), \mathscr{D}_{21}^*(\tau), \mathscr{D}^*(\tau)) \\
& \times \mathrm{vec}(\mathscr{F}_{11}(\tau, \mathscr{H}_{11}^*(\tau), K_x^*(\tau))) \\
& + \frac{\partial}{\partial \mathrm{vec}(\mathscr{Y}_{12})} \mathscr{W}(\tau, \mathscr{H}_{11}^*(\tau), \mathscr{H}_{12}^*(\tau), \mathscr{H}_{22}^*(\tau), \mathscr{D}_{11}^*(\tau), \mathscr{D}_{21}^*(\tau), \mathscr{D}^*(\tau)) \\
& \times \mathrm{vec}(\mathscr{F}_{12}(\tau, \mathscr{H}_{12}^*(\tau), \mathscr{H}_{11}^*(\tau), K_x^*(\tau), K_{z_1}^*(\tau))) \\
& + \frac{\partial}{\partial \mathrm{vec}(\mathscr{Y}_{22})} \mathscr{W}(\tau, \mathscr{H}_{11}^*(\tau), \mathscr{H}_{12}^*(\tau), \mathscr{H}_{22}^*(\tau), \mathscr{D}_{11}^*(\tau), \mathscr{D}_{21}^*(\tau), \mathscr{D}^*(\tau)) \\
& \times \mathrm{vec}(\mathscr{F}_{22}(\tau, \mathscr{H}_{22}^*(\tau), \mathscr{H}_{12}^*(\tau), K_{z_1}^*(\tau))) \\
& + \frac{\partial}{\partial \mathrm{vec}(\mathscr{Z}_{11})} \mathscr{W}(\tau, \mathscr{H}_{11}^*(\tau), \mathscr{H}_{12}^*(\tau), \mathscr{H}_{22}^*(\tau), \mathscr{Z}_{11}^*(\tau), \mathscr{Z}_{21}^*(\tau), \mathscr{D}^*(\tau)) \\
& \times \mathrm{vec}(\mathscr{G}_{11}(\tau, \mathscr{D}_{11}^*(\tau), \mathscr{H}_{11}^*(\tau), \mathscr{H}_{12}^*(\tau), K_x^*(\tau), v^*(\tau))) \\
& + \frac{\partial}{\partial \mathrm{vec}(\mathscr{Z}_{21})} \mathscr{W}(\tau, \mathscr{H}_{11}^*(\tau), \mathscr{H}_{12}^*(\tau), \mathscr{H}_{22}^*(\tau), \mathscr{D}_{11}^*(\tau), \mathscr{D}_{21}^*(\tau), \mathscr{D}^*(\tau)) \\
& \times \mathrm{vec}(\mathscr{G}_{21}(\tau, \mathscr{D}_{21}^*(\tau), \mathscr{D}_{11}^*(\tau), \mathscr{H}_{12}^*(\tau), \mathscr{H}_{22}^*(\tau), K_{z_1}^*(\tau), v^*(\tau))) \\
& + \frac{\partial}{\partial \mathrm{vec}(\mathscr{Z})} \mathscr{W}(\tau, \mathscr{H}_{11}^*(\tau), \mathscr{H}_{12}^*(\tau), \mathscr{H}_{22}^*(\tau), \mathscr{D}_{11}^*(\tau), \mathscr{D}_{21}^*(\tau), \mathscr{D}^*(\tau)) \\
& \times \mathrm{vec}(\mathscr{G}(\tau, \mathscr{H}_{11}^*(\tau), \mathscr{D}_{11}^*(\tau), \mathscr{D}_{21}^*(\tau), v^*(\tau)))
\end{aligned} \tag{6.51}$$

then the elements of K_x^*, $K_{z_1}^*$ *and* v^* *are optimal. Moreover, there holds*

$$\mathscr{W}(\varepsilon, \mathscr{Y}_{11}, \mathscr{Y}_{12}, \mathscr{Y}_{22}, \mathscr{Z}_{11}, \mathscr{Z}_{21}, \mathscr{Z}) = \mathscr{V}(\varepsilon, \mathscr{Y}_{11}, \mathscr{Y}_{12}, \mathscr{Y}_{22}, \mathscr{Z}_{11}, \mathscr{Z}_{21}, \mathscr{Z}),$$

whereby the value function is $\mathscr{V}(\varepsilon, \mathscr{Y}_{11}, \mathscr{Y}_{12}, \mathscr{Y}_{22}, \mathscr{Z}_{11}, \mathscr{Z}_{21}, \mathscr{Z})$.

Proof. Mathematical analysis for the proof can be found from the previous work available at [4]. □

6.4 Output-Feedback Control Solution

In this section, an important use of the Mayer-form verification theorem of dynamic programming in [3] is for finding a family of extremal paths that meet specified terminal conditions. Within the framework of dynamic programming, the terminal time and states $(t_f, \mathscr{H}_{11}^f, \mathscr{H}_{12}^f, \mathscr{H}_{22}^f, \mathscr{D}_{11}^f, \mathscr{D}_{21}^f, \mathscr{D}^f)$ of the dynamics (6.43)–(6.48) are now parameterized as $(\varepsilon, \mathscr{Y}_{11}, \mathscr{Y}_{12}, \mathscr{Y}_{22}, \mathscr{Z}_{11}, \mathscr{Z}_{21}, \mathscr{Z})$ for a broader family of optimization problems.

For each interior point $(\varepsilon, \mathscr{Y}_{11}, \mathscr{Y}_{12}, \mathscr{Y}_{22}, \mathscr{Z}_{11}, \mathscr{Z}_{21}, \mathscr{Z}) \in \mathscr{Q}$, the continuously differential function associated with the value function satisfying the HJB equation (6.50) is sought of the form

$$\begin{aligned}
&\mathscr{W}(\varepsilon, \mathscr{Y}_{11}, \mathscr{Y}_{12}, \mathscr{Y}_{22}, \mathscr{Z}_{11}, \mathscr{Z}_{21}, \mathscr{Z}) \\
&= x_0^T \sum_{r=1}^{k} \mu_r (\mathscr{Y}_{11}^r + \mathscr{E}_{11}^r(\varepsilon)) x_0 + 2x_0^T \sum_{r=1}^{k} \mu_r (\mathscr{Y}_{12}^r + \mathscr{E}_{12}^r(\varepsilon)) z_{10} \\
&\quad + z_{10}^T \sum_{r=1}^{k} \mu_r (\mathscr{Y}_{22}^r + \mathscr{E}_{22}^r(\varepsilon)) z_{10} + 2x_0^T \sum_{r=1}^{k} \mu_r (\mathscr{Z}_{11}^r + \mathscr{T}_{11}^r(\varepsilon)) \\
&\quad + 2z_{10}^T \sum_{r=1}^{k} \mu_r (\mathscr{Z}_{21}^r + \mathscr{T}_{21}^r(\varepsilon)) + \sum_{r=1}^{k} \mu_r (\mathscr{Z}^r + \mathscr{T}^r(\varepsilon)),
\end{aligned} \tag{6.52}$$

whereby the parametric functions $\mathscr{E}_{11}^r \in \mathscr{C}^1([t_0,t_f];\mathbb{R}^n)$, $\mathscr{E}_{12}^r \in \mathscr{C}^1([t_0,t_f];\mathbb{R}^{n\times n_1})$, $\mathscr{E}_{22}^r \in \mathscr{C}^1([t_0,t_f];\mathbb{R}^{n_1})$, $\mathscr{T}_{11}^r \in \mathscr{C}^1([t_0,t_f];\mathbb{R}^n)$, $\mathscr{T}_{21}^r \in \mathscr{C}^1([t_0,t_f];\mathbb{R}^{n_1})$, and $\mathscr{T}^r \in \mathscr{C}^1([t_0,t_f];\mathbb{R})$ are to be determined.

Moreover, it can be shown that the derivative of $\mathscr{W}(\varepsilon, \mathscr{Y}_{11}, \mathscr{Y}_{12}, \mathscr{Y}_{22}, \mathscr{Z}_{11}, \mathscr{Z}_{21}, \mathscr{Z})$ with respect to time ε is

$$\begin{aligned}
&\frac{d}{d\varepsilon}\mathscr{W}(\varepsilon, \mathscr{Y}_{11}, \mathscr{Y}_{12}, \mathscr{Y}_{22}, \mathscr{Z}_{11}, \mathscr{Z}_{21}, \mathscr{Z}) = x_0^T \sum_{r=1}^{k} \mu_r \left(\mathscr{F}_{11}^r(\varepsilon, \mathscr{Y}_{11}, K_x) + \frac{d}{d\varepsilon}\mathscr{E}_{11}^r(\varepsilon) \right) x_0 \\
&\quad + 2x_0^T \sum_{r=1}^{k} \mu_r \left(\mathscr{F}_{12}^r(\varepsilon, \mathscr{Y}_{12}, \mathscr{Y}_{11}, K_x, K_{z_1}) + \frac{d}{d\varepsilon}\mathscr{E}_{12}^r(\varepsilon) \right) z_{10}
\end{aligned}$$

$$+z_{10}^T \sum_{r=1}^{k} \mu_r \left(\mathscr{F}_{22}^r(\varepsilon, \mathscr{Y}_{22}, \mathscr{Y}_{12}, K_{z_1}) + \frac{\mathrm{d}}{\mathrm{d}\varepsilon} \mathscr{E}_{22}^r(\varepsilon) \right) z_{10}$$

$$+2x_0^T \sum_{r=1}^{k} \mu_r \left(\mathscr{G}_{11}^r(\varepsilon, \mathscr{Z}_{11}, \mathscr{Y}_{11}, \mathscr{Y}_{12}, K_x, v) + \frac{\mathrm{d}}{\mathrm{d}\varepsilon} \mathscr{T}_{11}^r(\varepsilon) \right)$$

$$+2z_{10}^T \sum_{r=1}^{k} \mu_r \left(\mathscr{G}_{21}^r(\varepsilon, \mathscr{Z}_{21}, \mathscr{Z}_{11}, \mathscr{Y}_{12}, \mathscr{Y}_{22}, K_{z_1}, v) + \frac{\mathrm{d}}{\mathrm{d}\varepsilon} \mathscr{T}_{21}^r(\varepsilon) \right)$$

$$+\sum_{r=1}^{k} \mu_r \left(\mathscr{G}^r(\varepsilon, \mathscr{Y}_{11}, \mathscr{Z}_{11}, \mathscr{Z}_{21}, v) + \frac{\mathrm{d}}{\mathrm{d}\varepsilon} \mathscr{T}^r(\varepsilon) \right) \tag{6.53}$$

provided that $(K_x, K_{z_1}, v) \in \overline{J}_x \times \overline{J}_{z_1} \times \overline{V}$.

The substitution of this candidate for the value function into the Hamilton-Jacobi-Bellman equation (6.50) and making use of Eq. (6.53) yield

$$0 = \min_{(K_x, K_{z_1}, v) \in \overline{J}_x \times \overline{J}_{z_1} \times \overline{V}} \Big\{ x_0^T \sum_{r=1}^{k} \mu_r (\mathscr{F}_{11}^r(\varepsilon, \mathscr{Y}_{11}, K_x) + \frac{\mathrm{d}}{\mathrm{d}\varepsilon} \mathscr{E}_{11}^r(\varepsilon)) x_0$$

$$+2x_0^T \sum_{r=1}^{k} \mu_r (\mathscr{F}_{12}^r(\varepsilon, \mathscr{Y}_{12}, \mathscr{Y}_{11}, K_x, K_{z_1}) + \frac{\mathrm{d}}{\mathrm{d}\varepsilon} \mathscr{E}_{12}^r(\varepsilon)) z_{10}$$

$$+z_{10}^T \sum_{r=1}^{k} \mu_r (\mathscr{F}_{22}^r(\varepsilon, \mathscr{Y}_{22}, \mathscr{Y}_{12}, K_{z_1}) + \frac{\mathrm{d}}{\mathrm{d}\varepsilon} \mathscr{E}_{22}^r(\varepsilon)) z_{10}$$

$$+2x_0^T \sum_{r=1}^{k} \mu_r (\mathscr{G}_{11}^r(\varepsilon, \mathscr{Z}_{11}, \mathscr{Y}_{11}, \mathscr{Y}_{12}, K_x, v) + \frac{\mathrm{d}}{\mathrm{d}\varepsilon} \mathscr{T}_{11}^r(\varepsilon))$$

$$+2z_{10}^T \sum_{r=1}^{k} \mu_r (\mathscr{G}_{21}^r(\varepsilon, \mathscr{Z}_{21}, \mathscr{Z}_{11}, \mathscr{Y}_{12}, \mathscr{Y}_{22}, K_{z_1}, v) + \frac{\mathrm{d}}{\mathrm{d}\varepsilon} \mathscr{T}_{21}^r(\varepsilon))$$

$$+\sum_{r=1}^{k} \mu_r (\mathscr{G}^r(\varepsilon, \mathscr{Y}_{11}, \mathscr{Z}_{11}, \mathscr{Z}_{21}, v) + \frac{\mathrm{d}}{\mathrm{d}\varepsilon} \mathscr{T}^r(\varepsilon)) \Big\}. \tag{6.54}$$

Differentiating the expression within the bracket of Eq. (6.54) with respect to K_x, K_{z_1}, and v yields the necessary conditions for an extremum of the Mayer-type performance index (6.49) on the time interval $[t_0, \varepsilon]$:

$$K_x(\varepsilon, \mathscr{Y}_{11}) = -R^{-1}(\varepsilon) B^T(\varepsilon) \sum_{r=1}^{k} \hat{\mu}_r \mathscr{Y}_{11}^r, \tag{6.55}$$

$$K_{z_1}(\varepsilon, \mathscr{Y}_{12}) = -R^{-1}(\varepsilon) B^T(\varepsilon) \sum_{r=1}^{k} \hat{\mu}_r \mathscr{Y}_{12}^r, \tag{6.56}$$

$$v(\varepsilon, \mathscr{Z}_{11}) = -R^{-1}(\varepsilon) B^T(\varepsilon) \sum_{r=1}^{k} \hat{\mu}_r \mathscr{Z}_{11}^r, \tag{6.57}$$

where the normalized weights $\hat{\mu}_r = \mu_r / \mu_1$ with $\mu_1 > 0$.

The dynamical equations (6.43)–(6.48) evaluated at the feedback (6.55), the feedforward (6.56), and the affine input (6.57) become: the backward-in-time matrix differential equations

$$\frac{d}{d\varepsilon}\mathscr{H}_{11}^{1}(\varepsilon) = -[A(\varepsilon)+B(\varepsilon)K_x(\varepsilon)]^T \mathscr{H}_{11}^{1}(\varepsilon) - \mathscr{H}_{11}^{1}(\varepsilon)[A(\varepsilon)+B(\varepsilon)K_x(\varepsilon)] - C^T(\varepsilon)Q(\varepsilon)C(\varepsilon) - K_x^T(\varepsilon)R(\varepsilon)K_x(\varepsilon), \tag{6.58}$$

$$\frac{d}{d\varepsilon}\mathscr{H}_{11}^{r}(\varepsilon) = -[A(\varepsilon)+B(\varepsilon)K_x(\varepsilon)]^T \mathscr{H}_{11}^{r}(\varepsilon) - \mathscr{H}_{11}^{r}(\varepsilon)[A(\varepsilon)+B(\varepsilon)K_x(\varepsilon)] - \sum_{s=1}^{r-1}\frac{2r!}{s!(r-s)!}\mathscr{H}_{11}^{s}(\varepsilon)L(\varepsilon)WL^T(\varepsilon)\mathscr{H}_{11}^{(r-s)}(\varepsilon) - \sum_{s=1}^{r-1}\frac{2r!}{s!(r-s)!}\mathscr{H}_{12}^{s}(\varepsilon)L_1(\varepsilon)W_1L_1^T(\varepsilon)(\mathscr{H}_{12}^{(r-s)})^T(\varepsilon), \tag{6.59}$$

$$\frac{d}{d\varepsilon}\mathscr{H}_{12}^{1}(\varepsilon) = -[A(\varepsilon)+B(\varepsilon)K_x(\varepsilon)]^T \mathscr{H}_{12}^{1}(\varepsilon) - \mathscr{H}_{11}^{1}(\varepsilon)B(\varepsilon)K_{z_1}(\varepsilon) - \mathscr{H}_{12}^{1}(\varepsilon)A_1(\varepsilon) + C^T(\varepsilon)Q(\varepsilon)C_1(\varepsilon) - K_x^T(\varepsilon)R(\varepsilon)K_{z_1}(\varepsilon), \tag{6.60}$$

$$\frac{d}{d\varepsilon}\mathscr{H}_{12}^{r}(\varepsilon) = -[A(\varepsilon)+B(\varepsilon)K_x(\varepsilon)]^T \mathscr{H}_{12}^{r}(\varepsilon) - \mathscr{H}_{11}^{r}(\varepsilon)B(\varepsilon)K_{z_1}(\varepsilon) - \mathscr{H}_{12}^{r}(\varepsilon)A_1(\varepsilon) - \sum_{s=1}^{r-1}\frac{2r!}{s!(r-s)!}\mathscr{H}_{11}^{s}(\varepsilon)L(\tau)WL^T(\varepsilon)\mathscr{H}_{12}^{(r-s)}(\varepsilon) - \sum_{s=1}^{r-1}\frac{2r!}{s!(r-s)!}\mathscr{H}_{12}^{s}(\varepsilon)L_1(\tau)W_1L_1^T(\varepsilon)\mathscr{H}_{22}^{(r-s)}(\varepsilon), \tag{6.61}$$

$$\frac{d}{d\varepsilon}\mathscr{H}_{22}^{1}(\varepsilon) = -A_1^T(\varepsilon)\mathscr{H}_{22}^{1}(\varepsilon) - \mathscr{H}_{22}^{1}(\varepsilon)A_1(\varepsilon) - K_{z_1}^T(\varepsilon)B^T(\varepsilon)\mathscr{H}_{12}^{1}(\varepsilon) - (\mathscr{H}_{12}^{1})^T(\varepsilon)B(\varepsilon)K_{z_1}(\varepsilon) - C_1^T(\varepsilon)Q(\varepsilon)C_1(\varepsilon) - K_{z_1}^T(\varepsilon)R(\varepsilon)K_{z_1}(\varepsilon), \tag{6.62}$$

$$\frac{d}{d\varepsilon}\mathscr{H}_{22}^{r}(\varepsilon) = -A_1^T(\varepsilon)\mathscr{H}_{22}^{r}(\varepsilon) - \mathscr{H}_{22}^{r}(\varepsilon)A_1(\varepsilon) - (\mathscr{H}_{12}^{r})^T(\varepsilon)B(\varepsilon)K_{z_1}(\varepsilon) - K_{z_1}^T(\varepsilon)B^T(\varepsilon)\mathscr{H}_{12}^{r}(\varepsilon) - \sum_{s=1}^{r-1}\frac{2r!}{s!(r-s)!}(\mathscr{H}_{12}^{s})^T(\varepsilon)L(\varepsilon)WL^T(\varepsilon)\mathscr{H}_{12}^{r-s}(\varepsilon) - \sum_{s=1}^{r-1}\frac{2r!}{s!(r-s)!}\mathscr{H}_{22}^{s}(\varepsilon)L_1(\varepsilon)W_1L_1^T(\varepsilon)\mathscr{H}_{22}^{r-s}(\varepsilon) \tag{6.63}$$

and the backward-in-time vector differential equations

$$\frac{d}{d\varepsilon}\mathcal{D}_{11}^{1}(\varepsilon) = -\left[A(\varepsilon)+B(\tau)K_x(\varepsilon)\right]^T \mathcal{D}_{11}^{1}(\varepsilon) - \mathcal{H}_{11}^{1}(\varepsilon)B(\varepsilon)v(\varepsilon) - \mathcal{H}_{12}^{1}(\varepsilon)B_1(\varepsilon)u_r(\varepsilon) - K_x^T(\varepsilon)R(\varepsilon)v(\varepsilon), \quad (6.64)$$

$$\frac{d}{d\varepsilon}\mathcal{D}_{11}^{r}(\varepsilon) = -\left[A(\varepsilon)+B(\varepsilon)K_x(\varepsilon)\right]^T \mathcal{D}_{11}^{r}(\varepsilon) - \mathcal{H}_{11}^{r}(\varepsilon)B(\varepsilon)v(\varepsilon) - \mathcal{H}_{12}^{r}(\varepsilon)B_1(\varepsilon)u_r(\varepsilon), \quad (6.65)$$

$$\frac{d}{d\varepsilon}\mathcal{D}_{21}^{1}(\varepsilon) = -K_{z_1}^T(\varepsilon)B^T(\varepsilon)\mathcal{D}_{11}^{1}(\varepsilon) - (\mathcal{H}_{12}^{1})^T(\varepsilon)B(\varepsilon)v(\varepsilon) - A_1^T(\varepsilon)\mathcal{D}_{21}^{1}(\varepsilon) - \mathcal{H}_{22}^{1}(\varepsilon)B_1(\varepsilon)u_r(\varepsilon) - K_{z_1}^T(\varepsilon)R(\varepsilon)v(\varepsilon), \quad (6.66)$$

$$\frac{d}{d\varepsilon}\mathcal{D}_{21}^{r}(\varepsilon) = -K_{z_1}^T(\varepsilon)B^T(\varepsilon)\mathcal{D}_{11}^{r}(\varepsilon) - (\mathcal{H}_{12}^{r})^T(\varepsilon)B(\tau)v(\varepsilon) - A_1^T(\varepsilon)\mathcal{D}_{21}^{r}(\varepsilon) - \mathcal{H}_{22}^{r}(\varepsilon)B_1(\varepsilon)u_r(\varepsilon) \quad (6.67)$$

and the backward-in-time scalar differential equations

$$\frac{d}{d\varepsilon}\mathcal{D}^{1}(\varepsilon) = -\mathrm{Tr}\left\{\mathcal{H}_{11}^{1}(\varepsilon)L(\varepsilon)WL^T(\varepsilon)\right\} - \mathrm{Tr}\left\{\mathcal{H}_{22}^{1}(\varepsilon)L_1(\varepsilon)W_1L_1^T(\varepsilon)\right\} - 2(\mathcal{D}_{11}^{1})^T(\varepsilon)B(\varepsilon)v(\varepsilon) - 2(\mathcal{D}_{21}^{1})^T(\varepsilon)B_1(\varepsilon)u_r(\varepsilon) - v^T(\varepsilon)R(\varepsilon)v(\varepsilon), \quad (6.68)$$

$$\frac{d}{d\varepsilon}\mathcal{D}^{r}(\varepsilon) = -\mathrm{Tr}\left\{\mathcal{H}_{11}^{r}(\varepsilon)L(\varepsilon)WL^T(\varepsilon)\right\} - \mathrm{Tr}\left\{\mathcal{H}_{22}^{r}(\varepsilon)L_1(\varepsilon)W_1L_1^T(\varepsilon)\right\} - 2(\mathcal{D}_{11}^{r})^T(\varepsilon)B(\varepsilon)v(\varepsilon) - 2(\mathcal{D}_{21}^{r})^T(\varepsilon)B_1(\varepsilon)u_r(\varepsilon), \quad (6.69)$$

whereby the terminal-value conditions are $\mathcal{H}_{11}^{1}(t_f) = C^T(t_f)Q_fC(t_f)$, $\mathcal{H}_{11}^{r}(t_f) = 0$ for $2 \le r \le k$; $\mathcal{H}_{12}^{1}(t_f) = -C^T(t_f)Q_fC_1(t_f)$, $\mathcal{H}_{12}^{r}(t_f) = 0$ for $2 \le r \le k$; $\mathcal{H}_{22}^{1}(t_f) = C_1^T(t_f)Q_fC_1(t_f)$, $\mathcal{H}_{22}^{r}(t_f) = 0$ for $2 \le r \le k$; $\mathcal{D}_{11}^{r}(t_f) = 0$ for $1 \le r \le k$; $\mathcal{D}_{21}^{r}(t_f) = 0$ for $1 \le r \le k$; $\mathcal{D}^{r}(t_f) = 0$ for $1 \le r \le k$.

To have the minimum of Eq. (6.54) equal to zero for any $\varepsilon \in [t_0, t_f]$ and when $\mathcal{Y}_{11}^{r}$, $\mathcal{Y}_{12}^{r}$, $\mathcal{Y}_{22}^{r}$, $\mathcal{Z}_{11}^{r}$, $\mathcal{Z}_{21}^{r}$ and $\mathcal{Z}^{r}$ evaluated along the solutions to Eqs. (6.43)–(6.48), the time-dependent functions $\mathcal{E}_{11}^{r}(\varepsilon)$, $\mathcal{E}_{12}^{r}(\varepsilon)$, $\mathcal{E}_{22}^{r}(\varepsilon)$, $\mathcal{T}_{11}^{r}(\varepsilon)$, $\mathcal{T}_{21}^{r}(\varepsilon)$ and $\mathcal{T}^{r}(\varepsilon)$ must satisfy the following differential equations

$$\frac{d}{d\varepsilon}\mathcal{E}_{11}^{r}(\varepsilon) = -\frac{d}{d\varepsilon}\mathcal{H}_{11}^{r}(\varepsilon), \qquad \frac{d}{d\varepsilon}\mathcal{E}_{12}^{r}(\varepsilon) = -\frac{d}{d\varepsilon}\mathcal{H}_{12}^{r}(\varepsilon), \quad (6.70)$$

$$\frac{d}{d\varepsilon}\mathcal{E}_{22}^{r}(\varepsilon) = -\frac{d}{d\varepsilon}\mathcal{H}_{22}^{r}(\varepsilon), \qquad \frac{d}{d\varepsilon}\mathcal{T}_{11}^{r}(\varepsilon) = -\frac{d}{d\varepsilon}\mathcal{D}_{11}^{r}(\varepsilon), \quad (6.71)$$

$$\frac{d}{d\varepsilon}\mathcal{T}_{21}^{r}(\varepsilon) = -\frac{d}{d\varepsilon}\mathcal{D}_{21}^{r}(\varepsilon), \qquad \frac{d}{d\varepsilon}\mathcal{T}^{r}(\varepsilon) = -\frac{d}{d\varepsilon}\mathcal{D}^{r}(\varepsilon). \quad (6.72)$$

The initial-value conditions for Eqs. (6.70)–(6.72) are then obtained, e.g., $\mathscr{E}_{11}^r(t_0) = 0$, $\mathscr{E}_{12}^r(t_0) = 0$, $\mathscr{E}_{22}^r(t_0) = 0$, $\mathscr{T}_{11}^r(t_0) = 0$, $\mathscr{T}_{21}^r(t_0) = 0$, and $\mathscr{T}^r(t_0) = 0$ for all $1 \leq r \leq k$.

Therefore, the sufficient condition (6.51) of the verification theorem is finally satisfied so the extremizing feedback gain (6.55), feedforward gain (6.56) and affine input (6.57) are optimal

$$K_x^*(\varepsilon) = -R^{-1}(\varepsilon)B^T(\varepsilon)\sum_{r=1}^{k}\widehat{\mu}_r\mathscr{H}_{11}^{r*}(\varepsilon), \tag{6.73}$$

$$K_{z_1}^*(\varepsilon) = -R^{-1}(\varepsilon)B^T(\varepsilon)\sum_{r=1}^{k}\widehat{\mu}_r\mathscr{H}_{12}^{r*}(\varepsilon), \tag{6.74}$$

$$v^*(\varepsilon) = -R^{-1}(\varepsilon)B^T(\varepsilon)\sum_{r=1}^{k}\widehat{\mu}_r\mathscr{D}_{11}^{r*}(\varepsilon), \tag{6.75}$$

whereby the optimal matrix-valued solutions $\{\mathscr{H}_{11}^{r*}(\cdot)\}_{r=1}^k$, $\{\mathscr{H}_{12}^{r*}(\cdot)\}_{r=1}^k$ and the optimal vector-valued solutions $\{\mathscr{D}_{11}^{r*}(\cdot)\}_{r=1}^k$ are solving the backward-in-time differential equations (6.58)–(6.59), (6.60)–(6.61), and (6.64)–(6.65), respectively.

6.5 Chapter Summary

The emphasis herein is placed on the analytical method for the design of output constrained feedback controller capable of compensating for stochastic model-following linear systems with risk-value aware performance index. There are tight couplings between control and estimation. Therefore, the "separation principle" does not apply in a general technique for the solution with optimal output feedback. The risk-averse control is of three-degrees-of-freedom type composed by the feedback gain, the feedforward and the affine input. In addition, the feedback gain is independent of the system parameters of reference and command-generating models.

References

1. Davison, E.J.: The feedforward control of linear multivariable time-invariant systems. Automatica **9**, 561–573 (1973)
2. Davison, E.J.: The steady-state invertibility and feedforward control of linear time-invariant systems. IEEE Trans. Automat. Contr. **21**, 529–534 (1976)
3. Fleming, W.H., Rishel, R.W.: Deterministic and Stochastic Optimal Control. Springer, New York (1975)

4. Pham, K.D.: Statistical control paradigms for structural vibration suppression. Ph.D. Dissertation. Department of Electrical Engineering. University of Notre Dame, Indiana, U.S.A. Available via http://etd.nd.edu/ETD-db/theses/available/etd-04152004-121926/unrestricted/PhamKD052004.pdf. Cited 15 March 2012 (2004)
5. Pham, K.D.: On statistical control of stochastic servo-systems: performance-measure statistics and state-feedback paradigm. In: Proceedings of the 17th International Federation of Automatic Control World Congress, pp. 7994–8000 (2008)
6. Tse, E.: On the optimal control of stochastic linear systems. IEEE Trans. Automat. Contr. **16**(6), 776–785 (1971)
7. Yuksel, S., Hindi, H., Crawford, L.: Optimal tracking with feedback-feedforward control separation over a network. In: Proceedings of the American Control Conference, pp. 3500–3506 (2006)

Chapter 7
Reliable Control for Stochastic Systems with Low Sensitivity

Abstract In this chapter, the problem of controlling stochastic linear systems with quadratic criterion which includes sensitivity variables is investigated. It is proved that the optimal full state-feedback control law with risk aversion can be realized by the cascade of mathematical statistics of performance uncertainty and a linear feedback. A set of nonlinear matrix equations are obtained, which constitutes the necessary and sufficient conditions that must be satisfied for an optimal solution.

7.1 Introduction

Here, as in the traditional approaches to trajectory and performance index sensitivity reductions, the main objective has been achieving a trade-off between optimality in the nominal performance and sensitivity to small parameter variations. In fact, discussions on open-loop and feedback control for deterministic linear regulators in [1, 2] centered largely on the inclusion of a quadratic trajectory sensitivity term in the integrand of the performance index.

Following the order of presentations on parameter sensitivity reduction in linear regulators for deterministic cases [1, 2], there is an emerging need of explicit recognition of a unified and systematic description of analysis and decision problems within a wider class of stochastic systems, described by higher-order characteristics of performance uncertainty and by controller designs with performance risk aversion. If the concept of the so-called statistical optimal control and its applications to analysis and control synthesis for a class of stochastic systems subject to linear dynamics, quadratic cost functionals, and additive Wiener noise processes corrupting the system dynamics and measurements developed in recent years [3, 4] is considered, then an attempt at a uniform theory of statistical optimal control with low sensitivity including considerations of combined trajectory and performance index sensitivity reduction is the main aim of the research investigation herein.

K.D. Pham, *Linear-Quadratic Controls in Risk-Averse Decision Making*, SpringerBriefs in Optimization, DOI 10.1007/978-1-4614-5079-5_7,

The structure of this chapter is as follows. Section 7.2 contains the control problem and the notion of admissible control. In addition, the development of all the mathematical statistics for performance robustness is carefully discussed. The detailed problem statements and solution method in obtaining the stochastic optimal control are described in Sects. 7.3 and 7.4. In Sect. 7.5, some conclusions and future research extension are provided.

7.2 The Problem and Representations for Performance Robustness

In this section, some classes of control problems with complete information and trajectory sensitivity considerations shall be investigated. For instance, the stochastic system being controlled on $[t_0,t_f]$ is linear and depends on the constant parameter variable ζ with the known initial condition $x(t_0) = x_0$

$$dx(t) = (A(t,\zeta)x(t) + B(t,\zeta)u(t))dt + G(t)dw(t), \tag{7.1}$$

whereby the continuous-time coefficients $A \in \mathscr{C}([t_0,t_f];\mathbb{R}^{n\times n})$, $B \in \mathscr{C}([t_0,t_f];\mathbb{R}^{n\times m})$ are the functions of the constant parameter variable ζ, except $G \in \mathscr{C}([t_0,t_f];\mathbb{R}^{n\times p})$. The process noise $w(t) \in \mathbb{R}^p$ is the p-dimensional stationary Wiener process defined on some complete probability space $(\Omega, \mathscr{F}, \mathscr{P})$ adapted over $[t_0,t_f]$ with the correlation of increments for all $\tau_1, \tau_2 \in [t_0,t_f]$ and $W > 0$

$$E\left\{[w(\tau_1) - w(\tau_2)][w(\tau_1) - w(\tau_2)]^T\right\} = W|\tau_1 - \tau_2|.$$

Notice that the state $x(t)$ is considered as the function of the constant parameter variable ζ, i.e., $x(t) \equiv \pi(t,x_0,\zeta)$. The traditional approach to sensitivity analysis for the system trajectory has been to define a sensitivity variable by

$$\sigma(t) \triangleq \frac{\partial}{\partial\zeta}\pi(t,x_0,\zeta)\bigg|_{\zeta=\zeta_0}. \tag{7.2}$$

Further on, the performance appraisal for controlling the system (7.1) is a finite-horizon integral-quadratic-form (IQF) random cost $J : \mathscr{C}([t_0,t_f];\mathbb{R}^m) \mapsto \mathbb{R}^+$:

$$J(u(\cdot)) = x^T(t_f)Q_f x(t_f) + \int_{t_0}^{t_f} [x^T(\tau)Q(\tau)x(\tau) + u^T(\tau)R(\tau)u(\tau) + \sigma^T(\tau)S(\tau)\sigma(\tau)]d\tau \tag{7.3}$$

wherein the symmetric state and sensitivity weightings Q and $S \in \mathscr{C}([t_0,t_f];\mathbb{R}^{n\times n})$ are positive semidefinite as is the terminal penalty weighting symmetric matrix $Q_f \in \mathbb{R}^{n\times n}$. The control effort weighting symmetric matrix $R \in \mathscr{C}([t_0,t_f];\mathbb{R}^{m\times m})$ is positive definite.

Next, the notion of admissible controls is discussed. In the case of complete information (e.g., state-feedback measurement), an admissible control is therefore of the form

$$u(t) = \gamma(t, x(t)), \quad t \in [t_0, t_f].$$

As shown in [3], the search for optimal control solutions to the statistical optimal control problem is consistently and productively restricted to linear time-varying feedback laws generated from the state $x(t)$ by

$$u(t) = K(t)x(t), \quad t \in [t_0, t_f] \tag{7.4}$$

with $K \in \mathscr{C}([t_0, t_f]; \mathbb{R}^{m \times n})$ an admissible gain whose further defining properties will be stated shortly.

Then, for an admissible $K(\cdot)$ and the admissible pair (t_0, x_0), it gives a sufficient condition for the existence of $x(t)$ in Eq. (7.1). In view of the control law (7.4), the controlled system (7.1) with the initial condition $x(t_0) = x_0$ is rewritten as follows:

$$\mathrm{d}x(t) = (A(t,\zeta) + B(t,\zeta)K(t))x(t)\mathrm{d}t + G(t)\mathrm{d}w(t). \tag{7.5}$$

In addition, the variation of the parameter ζ from the nominal value ζ_0 was assumed to be small, such that the Taylor's expansion of the state and control functions can be approximated by retaining only the first two terms. Differentiating the controlled diffusion process (7.5) with respect to ζ and evaluating at $\zeta = \zeta_0$ give

$$\begin{aligned}\mathrm{d}\sigma(t) &= (A_\zeta(t,\zeta_0) + B_\zeta(t,\zeta_0)K(t))x(t)\mathrm{d}t \\ &\quad + (A(t,\zeta_0) + B(t,\zeta_0)K(t))\sigma(t)\mathrm{d}t, \quad \sigma(t_0) = 0,\end{aligned} \tag{7.6}$$

whereby the parameter-dependent coefficients $A_\zeta \triangleq \frac{\partial A}{\partial \zeta}\Big|_{\zeta=\zeta_0}$ and $B_\zeta \triangleq \frac{\partial B}{\partial \zeta}\Big|_{\zeta=\zeta_0}$.

Henceforth, for the admissible pair (t_0, x_0) and control policy (7.4), the aggregation of Eqs. (7.5) and (7.6) is described by the controlled stochastic differential equation

$$\mathrm{d}z(t) = F_a(t)z(t)\mathrm{d}t + G_a(t)\mathrm{d}w(t), \quad z(t_0) = z_0 \tag{7.7}$$

with the performance measure (7.3) rewritten accordingly

$$J(K(\cdot)) = z^T(t_f)N_f z(t_f) + \int_{t_0}^{t_f} z^T(\tau)N(\tau)z(\tau)\mathrm{d}\tau, \tag{7.8}$$

whereby the aggregate state variables $z = \begin{bmatrix} x^T & \sigma^T \end{bmatrix}^T$ and the continuous-time coefficients are given by

$$F_a(t) = \begin{bmatrix} A(t,\zeta) + B(t,\zeta)K(t) & 0 \\ A_\zeta(t,\zeta_0) + B_\zeta(t,\zeta_0)K(t) & A(t,\zeta_0) + B(t,\zeta_0)K(t) \end{bmatrix}, \quad G_a(t) = \begin{bmatrix} G(t) \\ 0 \end{bmatrix},$$

$$N_f = \begin{bmatrix} Q_f & 0 \\ 0 & 0 \end{bmatrix}, \quad N(t) = \begin{bmatrix} Q(t) + K^T(t)R(t)K(t) & 0 \\ 0 & S(t) \end{bmatrix}, \quad z_0 = \begin{bmatrix} x_0 \\ 0 \end{bmatrix}.$$

In the analysis and design of performance-based uncertain systems, it is important to investigate a relation concerning two different subjects of knowledge, e.g,. process information (7.7) and goal information (7.8). In the descriptive approach herein, the control designer is concerned with performance variations described by knowledge representation in the form of a complete probabilistic description of the generalized chi-squared random variable (7.8). The learning process consists here in step by step knowledge validation and updating. At each step all the mathematical statistics associated with Eq. (7.8) in the knowledge updating should be determined before knowledge validation. The results of the successive determination of these performance-measure statistics associated with Eq. (7.8) are used in the formulation of the decisions in a learning decision making with performance risk aversion.

Mathematically stated, the knowledge validation and updating about performance variations of Eq. (7.8) is concerned with modeling and management of cost cumulants (also known as semi-invariants) associated with Eq. (7.8) as shown below.

Theorem 7.2.1 (Cumulant-Generating Function). *Let $z(\cdot)$ be a state variable of the stochastic dynamics concerning sensitivity (7.7) with initial values $z(\tau) \equiv z_\tau$ and $\tau \in [t_0, t_f]$. Further, let the moment-generating function be denoted by*

$$\varphi(\tau, z_\tau; \theta) = \rho(\tau, \theta) \exp\{z_\tau^T \Upsilon(\tau, \theta) z_\tau\}, \tag{7.9}$$

$$\upsilon(\tau, \theta) = \ln\{\rho(\tau, \theta)\}, \qquad \theta \in \mathbb{R}^+. \tag{7.10}$$

Then, the cumulant-generating function has the form of quadratic affine

$$\psi(\tau, z_\tau; \theta) = z_\tau^T \Upsilon(\tau, \theta) z_\tau + \upsilon(\tau, \theta), \tag{7.11}$$

where the scalar solution $\upsilon(\tau, \theta)$ solves the backward-in-time differential equation

$$\frac{d}{d\tau} \upsilon(\tau, \theta) = -\mathrm{Tr}\left\{\Upsilon(\tau, \theta) G_a(\tau) W G_a^T(\tau)\right\}, \quad \upsilon(t_f, \theta) = 0 \tag{7.12}$$

and the matrix solution $\Upsilon(\tau, \theta)$ satisfies the backward-in-time differential equation

$$\begin{aligned} \frac{d}{d\tau} \Upsilon(\tau, \theta) &= -F_a^T(\tau) \Upsilon(\tau, \theta) - \Upsilon(\tau, \theta) F_a(\tau) \\ &\quad -2\Upsilon(\tau, \theta) G_a(\tau) W G_a^T(\tau) \Upsilon(\tau, \theta) - \theta N(\tau), \quad \Upsilon(t_f, \theta) = \theta N_f. \end{aligned} \tag{7.13}$$

Meanwhile, the solution $\rho(\tau,\theta)$ satisfies the time-backward differential equation

$$\frac{d}{d\tau}\rho(\tau,\theta) = -\rho(\tau,\theta)\,\mathrm{Tr}\{\Upsilon(\tau,\theta)G_a(\tau)WG_a^T(\tau)\}, \quad \rho(t_f,\theta) = 1. \tag{7.14}$$

Proof. For notional simplicity, it is convenient to have $\varpi(\tau,z_\tau;\theta) \triangleq \exp\{\theta J(\tau,z_\tau)\}$ in which the performance measure (7.8) is rewritten as the cost-to-go function from an arbitrary state z_τ at a running time $\tau \in [t_0,t_f]$, that is,

$$J(\tau,z_\tau) = z^T(t_f)N_f z(t_f) + \int_\tau^{t_f} z^T(t)N(t)z(t)dt \tag{7.15}$$

subject to

$$dz(t) = F_a(t)z(t)dt + G_a(t)dw(t), \qquad z(\tau) = z_\tau. \tag{7.16}$$

By definition, the moment-generating function is $\varphi(\tau,z_\tau;\theta) \triangleq E\{\varpi(\tau,z_\tau;\theta)\}$. Thus, the total time derivative of $\varphi(\tau,z_\tau;\theta)$ is obtained as

$$\frac{d}{d\tau}\varphi(\tau,z_\tau;\theta) = -\theta z_\tau^T N(\tau)z_\tau\varphi(\tau,z_\tau;\theta).$$

Using the standard Ito's formula, it follows

$$\begin{aligned}
d\varphi(\tau,z_\tau;\theta) &= E\{d\varpi(\tau,z_\tau;\theta)\} \\
&= E\Big\{\varpi_\tau(\tau,z_\tau;\theta)d\tau + \varpi_{z_\tau}(\tau,z_\tau;\theta)dz_\tau \\
&\quad + \frac{1}{2}\mathrm{Tr}\{\varpi_{z_\tau z_\tau}(\tau,z_\tau;\theta)G_a(\tau)WG_a^T(\tau)\}d\tau\Big\} \\
&= \varphi_\tau(\tau,z_\tau;\theta)d\tau + \varphi_{z_\tau}(\tau,z_\tau;\theta)F_a(\tau)z_\tau d\tau \\
&\quad + \frac{1}{2}\mathrm{Tr}\{\varphi_{z_\tau z_\tau}(\tau,z_\tau;\theta)G_a(\tau)WG_a^T(\tau)\}d\tau
\end{aligned}$$

which under the hypothesis of $\varphi(\tau,z_\tau;\theta) = \rho(\tau,\theta)\exp\{z_\tau^T\Upsilon(\tau,\theta)z_\tau\}$ and its partial derivatives leads to the result

$$\begin{aligned}
-\theta z_\tau^T N(\tau)z_\tau\varphi(\tau,z_\tau;\theta) &= \Bigg\{\frac{\frac{d}{d\tau}\rho(\tau,\theta)}{\rho(\tau,\theta)} + z_\tau^T\frac{d}{d\tau}\Upsilon(\tau,\theta)z_\tau + z_\tau^T[F_a^T(\tau)\Upsilon(\tau,\theta) \\
&\quad + \Upsilon(\tau,\theta)F_a(\tau) + 2\Upsilon(\tau,\theta)G_a(\tau)WG_a^T(\tau)\Upsilon(\tau,\theta)]z_\tau \\
&\quad + \mathrm{Tr}\{\Upsilon(\tau,\theta)G_a(\tau)WG_a^T(\tau)\}\Bigg\}\varphi(\tau,z_\tau;\theta).
\end{aligned}$$

To have constant and quadratic terms being independent of arbitrary z_τ, it requires

$$\frac{\mathrm{d}}{\mathrm{d}\tau}\Upsilon(\tau,\theta) = -F_a^T(\tau)\Upsilon(\tau,\theta) - \Upsilon(\tau,\theta)F_a(\tau) \\ - 2\Upsilon(\tau,\theta)G_a(\tau)WG_a^T(\tau)\Upsilon(\tau,\theta) - \theta N(\tau),$$

$$\frac{\mathrm{d}}{\mathrm{d}\tau}\rho(\tau,\theta) = -\rho(\tau,\theta)\,\mathrm{Tr}\left\{\Upsilon(\tau,\theta)G_a(\tau)WG_a^T(\tau)\right\}$$

with the terminal-value conditions $\Upsilon(t_f,\theta) = \theta N_f$ and $\rho(t_f,\theta) = 1$. Finally, the backward-in-time differential equation satisfied by $\upsilon(\tau,\theta)$ is obtained

$$\frac{\mathrm{d}}{\mathrm{d}\tau}\upsilon(\tau,\theta) = -\mathrm{Tr}\left\{\Upsilon(\tau,\theta)G_a(\tau)WG_a^T(\tau)\right\}, \quad \upsilon(t_f,\theta) = 0,$$

which completes the proof. □

As it turns out that all the higher-order characteristic distributions associated with performance uncertainty and risk are well captured in the higher-order performance-measure statistics associated with Eq. (7.8). Subsequently, higher-order statistics that encapsulate the uncertain nature of Eq. (7.8) can now be generated via a Maclaurin series expansion of the cumulant-generating function (7.11):

$$\psi(\tau,z_\tau;\theta) = \sum_{r=1}^{\infty} \frac{\partial^{(r)}}{\partial\theta^{(r)}}\psi(\tau,z_\tau;\theta)\bigg|_{\theta=0} \frac{\theta^r}{r!} \tag{7.17}$$

in which all $\kappa_r \triangleq \frac{\partial^{(r)}}{\partial\theta^{(r)}}\psi(\tau,z_\tau;\theta)\Big|_{\theta=0}$ are called rth-order performance-measure statistics. Moreover, the series expansion coefficients are computed by using the cumulant-generating function (7.11):

$$\frac{\partial^{(r)}}{\partial\theta^{(r)}}\psi(\tau,z_\tau;\theta)\bigg|_{\theta=0} = z_\tau^T \frac{\partial^{(r)}}{\partial\theta^{(r)}}\Upsilon(\tau,\theta)\bigg|_{\theta=0} z_\tau + \frac{\partial^{(r)}}{\partial\theta^{(r)}}\upsilon(\tau,\theta)\bigg|_{\theta=0}. \tag{7.18}$$

In view of the definition (7.17), the rth-order performance-measure statistic thus follows

$$\kappa_r = z_\tau^T \frac{\partial^{(r)}}{\partial\theta^{(r)}}\Upsilon(\tau,\theta)\bigg|_{\theta=0} z_\tau + \frac{\partial^{(r)}}{\partial\theta^{(r)}}\upsilon(\tau,\theta)\bigg|_{\theta=0}, \quad 1 \le r < \infty. \tag{7.19}$$

In this chapter, it will be often convenient to change the notations

$$H_r(\tau) \triangleq \frac{\partial^{(r)}\Upsilon(\tau,\theta)}{\partial\theta^{(r)}}\bigg|_{\theta=0}, \quad D_r(\tau) \triangleq \frac{\partial^{(r)}\upsilon(\tau,\theta)}{\partial\theta^{(r)}}\bigg|_{\theta=0} \tag{7.20}$$

so that the next procedure that is illustrated in the theorem below provides an effective and accurate capability for forecasting all the higher-order characteristics associated with performance uncertainty. Therefore, via performance-measure statistics and adaptive decision making, it is anticipated that future performance variations will lose the element of surprise due to the inherent property of self-enforcing and risk-averse control solutions that are readily capable of reshaping the cumulative probability distribution of closed-loop performance.

Theorem 7.2.2 (Performance-Measure Statistics). *Let the linear-quadratic stochastic system be described by Eqs. (7.7)–(7.8), in which the pair (A,B) is uniformly stabilizable. For $k \in \mathbb{Z}^+$ fixed, the kth-order performance-measure statistic of the chi-squared random performance (7.8) is given by*

$$\kappa_k = z_0^T H_k(t_0) z_0 + D_k(t_0), \tag{7.21}$$

where the supporting variables $\{H_r(\tau)\}_{r=1}^k$ and $\{D_r(\tau)\}_{r=1}^k$ evaluated at $\tau = t_0$ satisfy the differential equations (with the dependence of $H_r(\tau)$ and $D_r(\tau)$ upon $K(\tau)$ suppressed)

$$\frac{d}{d\tau} H_1(\tau) = -F_a^T(\tau) H_1(\tau) - H_1(\tau) F_a(\tau) - N(\tau), \quad H_1(t_f) = N_f, \tag{7.22}$$

$$\frac{d}{d\tau} H_r(\tau) = -F_a^T(\tau) H_r(\tau) - H_r(\tau) F_a(\tau)$$
$$- \sum_{s=1}^{r-1} \frac{2r!}{s!(r-s)!} H_s(\tau) G_a(\tau) W G_a^T(\tau) H_{r-s}(\tau), \quad H_r(t_f) = 0, \tag{7.23}$$

$$\frac{d}{d\tau} D_r(\tau) = -\mathrm{Tr}\left\{ H_r(\tau) G_a(\tau) W G_a^T(\tau) \right\}, \quad D_r(t_f) = 0, \quad 1 \le r \le k. \tag{7.24}$$

Proof. The expression of performance-measure statistics described in Eq. (7.21) is readily justified by using result (7.19) and definition (7.20). What remains is to show that the solutions $H_r(\tau)$ and $D_r(\tau)$ for $1 \le r \le k$ indeed satisfy the dynamical equations (7.22)–(7.24). Notice that these backward-in-time equations (7.22)–(7.24) satisfied by $H_r(\tau)$ and $D_r(\tau)$ are therefore obtained by successively taking derivatives with respect to θ of the supporting equations (7.12)–(7.14) under the assumption of (A,B) uniformly stabilizable on $[t_0, t_f]$. □

To anticipate for a well-posed optimization problem, some sufficient conditions for the existence of solutions to the cumulant-generating equations (7.22)–(7.24) are now presented.

Theorem 7.2.3 (Existence of Performance-Measure Statistics). *Let (A,B) be uniformly stabilizable. Then, any given $k \in \mathbb{N}$, the backward-in-time matrix-valued differential equations (7.22)–(7.24) admit unique and bounded solutions $\{H_r(\tau)\}_{r=1}^k$ and $\{D_r(\tau)\}_{r=1}^k$ on $[t_0, t_f]$.*

Proof. Under the assumption of stabilizability, there always exists a feedback control gain such that the continuous-time composite state matrix $F_a(t)$ is uniformly exponentially stable on $[t_0,t_f]$. In other words, there exist positive constants η_1 and η_2 such that the pointwise matrix norm of the closed-loop state transition matrix satisfies the inequality

$$||\Phi_a(t,\tau)|| \leq \eta_1 e^{-\eta_2(t-\tau)}, \qquad \forall\, t \geq \tau \geq t_0 .$$

According to the results in [6], the state transition matrix, $\Phi_a(t,t_0)$ associated with the continuous-time composite state matrix $F_a(t)$ has the sequel properties

$$\frac{d}{dt}\Phi_a(t,t_0) = F_a(t)\Phi_a(t,t_0); \qquad \Phi_a(t_0,t_0) = I$$

$$\lim_{t_f\to\infty} ||\Phi_a(t_f,t)|| = 0; \quad \lim_{t_f\to\infty}\int_{t_0}^{t_f} ||\Phi_a(t_f,t)||^2 dt < \infty.$$

By the matrix variation of constant formula, the unique solutions to the time-backward matrix differential equations (7.22)–(7.24) together with the terminal-value conditions are then written as follows:

$$H_1(\tau) = \Phi_a^T(t_f,\tau)N_f\Phi(t_f,\tau) + \int_\tau^{t_f} \Phi_a^T(t,\tau)N(t)\Phi_a(t,\tau)dt,$$

$$H_r(\tau) = \int_\tau^{t_f} \Phi_a^T(t,\tau)\sum_{s=1}^{r-1}\frac{2r!}{s!(r-s)!}H_s(t)G_a(t)WG_a^T(t)H_{r-s}(t)\Phi_a(t,\tau)dt,$$

$$D_r(\tau) = \int_\tau^{t_f} \mathrm{Tr}\left\{H_r(t)G_a(t)WG_a^T(t)\right\}dt, \quad 1 \leq r \leq k.$$

As long as the growth rate of the integrals is not faster than the exponentially decreasing rates of two factors of $\Phi_a(\cdot,\tau)$, it is therefore concluded that there exist upper bounds on the solutions $H_r(\tau)$ and $D_r(\tau)$ during the finite interval $[t_0,t_f]$. □

7.3 Problem Statements with the Maximum Principle

The purpose of this section is to provide the statements of statistical optimal control with the addition of the necessary and sufficient conditions for optimality for risk-averse feedback strategies with low sensitivity that are considered in this investigation. The statistical optimal control of linear stochastic systems with low sensitivity here is distinguished by the fact that the evolution in time of all mathematical statistics (7.21) associated with the random performance measure (7.8) of the generalized chi-squared type is described by means of matrix-and scalar-valued differential equations (7.22)–(7.24).

For such problems, it is important to have a compact statement raised from statistical optimal control so as to aid mathematical manipulations. To make this more precise, one may think of the k-tuple state variables $\mathscr{H}(\cdot) \triangleq (\mathscr{H}_1(\cdot),\ldots,\mathscr{H}_k(\cdot))$ and $\mathscr{D}(\cdot) \triangleq (\mathscr{D}_1(\cdot),\ldots,\mathscr{D}_k(\cdot))$ whose continuously differentiable components $\mathscr{H}_r \in \mathscr{C}^1([t_0,t_f];\mathbb{R}^{2n\times 2n})$ and $\mathscr{D}_r \in \mathscr{C}^1([t_0,t_f];\mathbb{R})$ having the representations $\mathscr{H}_r(\cdot) \triangleq H_r(\cdot)$ and $\mathscr{D}_r(\cdot) \triangleq D_r(\cdot)$ with the right members satisfying the dynamics (7.22)–(7.24) are defined on the finite horizon $[t_0,t_f]$. In the remainder of the development, the bounded Lipschitz continuous mappings are introduced as

$$\mathscr{F}_r : [t_0,t_f] \times (\mathbb{R}^{2n\times 2n})^k \mapsto \mathbb{R}^{2n\times 2n},$$
$$\mathscr{G}_r : [t_0,t_f] \times (\mathbb{R}^{2n\times 2n})^k \mapsto \mathbb{R},$$

where the rules of action are given by

$$\mathscr{F}_1(\tau,\mathscr{H}) \triangleq -F_a^T(\tau)\mathscr{H}_1(\tau) - \mathscr{H}_1(\tau)F_a(\tau) - N(\tau),$$
$$\mathscr{F}_r(\tau,\mathscr{H}) \triangleq -F_a^T(\tau)\mathscr{H}_r(\tau) - \mathscr{H}_r(\tau)F_a(\tau)$$
$$-\sum_{s=1}^{r-1} \frac{2r!}{s!(r-s)!}\mathscr{H}_s(\tau)G_a(\tau)WG_a^T(\tau)\mathscr{H}_{r-s}(\tau), \quad 2 \le r \le k,$$
$$\mathscr{G}_r(\tau,\mathscr{H}) \triangleq -\mathrm{Tr}\left\{\mathscr{H}_r(\tau)G_a(\tau)WG_a^T(\tau)\right\}, \quad \le 1 \le r \le k.$$

The product mappings that follow are necessary for a compact formulation

$$\mathscr{F}_1 \times\cdots\times \mathscr{F}_k : [t_0,t_f] \times (\mathbb{R}^{2n\times 2n})^k \mapsto (\mathbb{R}^{2n\times 2n})^k,$$
$$\mathscr{G}_1 \times\cdots\times \mathscr{G}_k : [t_0,t_f] \times (\mathbb{R}^{2n\times 2n})^k \mapsto \mathbb{R}^k,$$

whereby the corresponding notations $\mathscr{F} \triangleq \mathscr{F}_1 \times\cdots\times \mathscr{F}_k$ and $\mathscr{G} \triangleq \mathscr{G}_1 \times\cdots\times \mathscr{G}_k$ are employed. Thus, the dynamic equations of motion (7.22)–(7.24) can be rewritten as

$$\frac{\mathrm{d}}{\mathrm{d}\tau}\mathscr{H}(\tau) = \mathscr{F}(\tau,\mathscr{H}(\tau)), \qquad \mathscr{H}(t_f) \equiv \mathscr{H}_f, \tag{7.25}$$
$$\frac{\mathrm{d}}{\mathrm{d}\tau}\mathscr{D}(\tau) = \mathscr{G}(\tau,\mathscr{H}(\tau)), \qquad \mathscr{D}(t_f) \equiv \mathscr{D}_f, \tag{7.26}$$

whereby the terminal-value conditions $\mathscr{H}_f \triangleq (N_f, 0, \ldots, 0)$ and $\mathscr{D}_f = (0,\ldots,0)$.

Notice that the product system uniquely determines the state matrices $\mathscr{H}$ and $\mathscr{D}$ once the admissible feedback gain K has been being specified. Henceforth, these state variables will be considered as $\mathscr{H} \equiv \mathscr{H}(\cdot,K)$ and $\mathscr{D} \equiv \mathscr{D}(\cdot,K)$. Therefore, the performance index in statistical optimal control problems is now formulated in K. For the given terminal data $(t_f,\mathscr{H}_f,\mathscr{D}_f)$, the classes of admissible feedback gains are next defined.

Definition 7.3.1 (Admissible Feedback Gains). Let compact subset $\overline{K} \subset \mathbb{R}^{m\times n}$ be the set of allowable feedback gain values. For the given $k \in \mathbb{Z}^+$ and sequence

$\mu = \{\mu_r \geq 0\}_{r=1}^k$ with $\mu_1 > 0$, the set of feedback gains $\mathscr{K}_{t_f,\mathscr{H}_f,\mathscr{D}_f;\mu}$ is assumed to be the class of $\mathscr{C}([t_0,t_f];\mathbb{R}^{m\times n})$ with values $K(\cdot) \in \overline{K}$ for which solutions to the dynamic equations (7.25)–(7.26) with the terminal-value conditions $\mathscr{H}(t_f) = \mathscr{H}_f$ and $\mathscr{D}(t_f) = \mathscr{D}_f$ exist on $[t_0,t_f]$ of optimization.

The development that follows is to present the application of uncertain performance variable to risk-averse decision making for the decision problem of the random variable (7.8) described by probability distribution of the generalized chi-squared type. The knowledge of decision making here is dealt by a new concept of risk-value aware performance index which naturally contains some tradeoffs between performance values and risks for the subject class of stochastic control problems.

On $\mathscr{K}_{t_f,\mathscr{H}_f,\mathscr{D}_f;\mu}$, the performance index with risk-value considerations in statistical optimal control is subsequently defined as follows:

Definition 7.3.2 (Risk-Value Aware Performance Index). Fix $k \in \mathbb{N}$ and the sequence of scalar coefficients $\mu = \{\mu_r \geq 0\}_{r=1}^k$ with $\mu_1 > 0$. Then, for the given z_0, the risk-value aware performance index

$$\phi_0 : \{t_0\} \times (\mathbb{R}^{2n\times 2n})^k \times \mathbb{R}^k \mapsto \mathbb{R}^+$$

pertaining to statistical optimal control of the stochastic system with low sensitivity over $[t_0,t_f]$ is defined by

$$\phi_0\left(t_0, \mathscr{H}(t_0), \mathscr{D}(t_0)\right) \triangleq \underbrace{\mu_1 \kappa_1}_{\text{Value Measure}} \quad \underbrace{+\mu_2\kappa_2 + \cdots + \mu_k\kappa_k}_{\text{Risk Measures}}$$

$$= \sum_{r=1}^{k} \mu_r \left[z_0^T \mathscr{H}_r(t_0) z_0 + \mathscr{D}_r(t_0)\right] \tag{7.27}$$

where additional design of freedom by means of μ_r's utilized by the control designer with risk-averse attitudes is sufficient to meet and exceed different levels of performance-based reliability requirements, for instance, mean (i.e., the average of performance measure), variance (i.e., the dispersion of values of performance measure around its mean), skewness (i.e., the antisymmetry of the probability density of performance measure), and kurtosis (i.e., the heaviness in the probability density tails of performance measure), pertaining to closed-loop performance variations and uncertainties while the supporting solutions $\{\mathscr{H}_r(\tau)\}_{r=1}^k$ and $\{\mathscr{D}_r(\tau)\}_{r=1}^k$ evaluated at $\tau = t_0$ satisfy the dynamical equations (7.25)–(7.26).

Next, the optimization statement over $[t_0,t_f]$ is stated.

Definition 7.3.3 (Optimization Problem of Mayer Type). Fix $k \in \mathbb{Z}^+$ and the sequence of scalar coefficients $\mu = \{\mu_r \geq 0\}_{r=1}^k$ with $\mu_1 > 0$. The optimization problem of statistical optimal control over $[t_0,t_f]$ is given by

$$\min_{K(\cdot)\in\mathscr{K}_{t_f,\mathscr{H}_f,\mathscr{D}_f;\mu}} \phi_0(t_0, \mathscr{H}(t_0), \mathscr{D}(t_0)) \tag{7.28}$$

subject to the dynamical equations (7.25)–(7.26) for $\tau \in [t_0, t_f]$.

Theorem 7.3.1 (Property of Risk-Value Performance Index). *If the compact and convex $\mathcal{K}_{t_f,\mathcal{H}_f,\mathcal{D}_f;\mu}$ is nonempty, then the performance index (7.27) is strictly convex in K.*

Proof. Indeed, the set of feedback strategies $\mathcal{K}_{t_f,\mathcal{H}_f,\mathcal{D}_f;\mu}$ is nonempty, compact, and convex. What remains is to show that the continuous function $\phi_0(t_0, \mathcal{H}(t_0), \mathcal{D}(t_0))$ is strictly convex in K. Such a case is illustrated by aggregating the dynamical equations (7.22)–(7.23), e.g.,

$$\frac{\mathrm{d}}{\mathrm{d}\tau}\Lambda(\tau) = -F_a^T(\tau)\Lambda(\tau) - \Lambda(\tau)F_a(\tau) - \mu_1 N(\tau)$$
$$- \sum_{r=2}^{k} \mu_r \sum_{s=1}^{r-1} \frac{2r!}{s!(r-s)!} \mathcal{H}_s(\tau)G_a(\tau)WG_a^T(\tau)\mathcal{H}_{r-s}(\tau), \qquad (7.29)$$

whereby the aggregate variable $\Lambda(\tau) \triangleq \sum_{r=1}^{k} \mu_r \mathcal{H}_r(\tau)$ and $\Lambda(t_f) = \mu_1 N_f$. The fundamental theorem of calculus and stochastic differential rule applied to the expression of $z^T(\tau)\Lambda(\tau)z(\tau)$ yield the result

$$E\left\{z^T(t_f)\mu_1 N_f z(t_f)\right\} - z_0^T \Lambda(t_0) z_0 = E\left\{\int_{t_0}^{t_f} d\left[z^T(\tau)\Lambda(\tau)z(\tau)\right]\right\}$$
$$= E\left\{\int_{t_0}^{t_f} [\mathrm{d}z^T(\tau)\Lambda(\tau)z(\tau) + z^T(\tau)\Lambda(\tau)\mathrm{d}z(\tau)\right.$$
$$\left. + z^T(\tau)\frac{\mathrm{d}}{\mathrm{d}\tau}\Lambda(\tau)z(\tau)\mathrm{d}\tau + \mathrm{d}z^T(\tau)\Lambda(\tau)\mathrm{d}z(\tau)]\right\}.$$

After some manipulations, it follows that

$$E\left\{z^T(t_f)\mu_1 N_f z(t_f)\right\} - z_0^T \Lambda(t_0) z_0 = \int_{t_0}^{t_f} \mathrm{Tr}\left\{\Lambda(\tau)G_a(\tau)WG_a^T(\tau)\right\}\mathrm{d}\tau$$
$$+ E\left\{\int_{t_0}^{t_f} z^T(\tau)[F_a(\tau)^T\Lambda(\tau) + \Lambda(\tau)F_a(\tau) + \frac{\mathrm{d}}{\mathrm{d}\tau}\Lambda(\tau)]z(\tau)\mathrm{d}\tau\right\}. \qquad (7.30)$$

Notice that the solution of Eq. (7.24) is written by an integral form

$$\mathcal{D}_r(t_0) = \int_{t_0}^{t_f} \mathrm{Tr}\left\{\mathcal{H}_r(\tau)G_a(\tau)WG_a^T(\tau)\right\}\mathrm{d}\tau, \quad 1 \le r \le k.$$

In view of the definition of $\Lambda(\cdot)$, it is then easy to see that

$$\sum_{r=1}^{k} \mu_r \mathcal{D}_r(t_0) = \int_{t_0}^{t_f} \mathrm{Tr}\left\{\Lambda(\tau)G_a(\tau)WG_a^T(\tau)\right\}\mathrm{d}\tau.$$

Henceforth, the performance index with risk consequences (7.27) is rewritten as

$$\phi_0(t_0, \mathcal{H}(t_0), \mathcal{D}(t_0)) = z_0^T \Lambda(t_0) z_0 + \int_{t_0}^{t_f} \mathrm{Tr}\left\{\Lambda(\tau) G_a(\tau) W G_a^T(\tau)\right\} \mathrm{d}\tau. \tag{7.31}$$

Replacing the results (7.29) and (7.31) into Eq. (7.30), it yields

$$\phi_0(t_0, \mathcal{H}(t_0), \mathcal{D}(t_0)) = E\left\{z^T(t_f)\mu_1 N_f z(t_f)\right\} + E\left\{\int_{t_0}^{t_f} z^T(\tau)\left[\mu_1 N(\tau) + \sum_{r=2}^{k}\mu_r \sum_{s=1}^{r-1} \frac{2r!}{s!(r-s)!} \mathcal{H}_s(\tau) G_a(\tau) W G_a^T(\tau) \mathcal{H}_{r-s}(\tau)\right] z(\tau) \mathrm{d}\tau\right\}, \tag{7.32}$$

which leads to

$$\phi_0(t_0, \mathcal{H}(t_0), \mathcal{D}(t_0)) = \mathrm{Tr}\left\{\mu_1 N_f P(t_f)\right\} + \mathrm{Tr}\left\{\int_{t_0}^{t_f} \left[\mu_1 N(\tau) + \sum_{r=2}^{k}\mu_r \sum_{s=1}^{r-1} \frac{2r!}{s!(r-s)!} \mathcal{H}_s(\tau) G_a(\tau) W G_a^T(\tau) \mathcal{H}_{r-s}(\tau)\right] P(\tau) \mathrm{d}\tau\right\}, \tag{7.33}$$

where the positive-definite solution $P(\cdot) \triangleq E\{z(\cdot)z^T(\cdot)\}$ is satisfying the forward-in-time matrix-valued differential equation

$$\frac{\mathrm{d}}{\mathrm{d}\tau} P(\tau) = P(\tau) F_a^T(\tau) + F_a(\tau) P(\tau) + G_a(\tau) W G_a^T(\tau), \quad P(t_0) = z_0 z_0^T. \tag{7.34}$$

Notice that within the integrand of Eq. (7.33), $N(\cdot)$ is strictly convex in K while other factors are positive semi-definite. Thus, the generalized performance index (7.33), or equivalently risk-value aware performance index (7.27) is strictly convex in K. □

7.4 Low Sensitivity Control with Risk Aversion

Opposite to the spirit of the earlier work [3, 4] relative to the traditional approach of dynamic programming to the optimization problem of Mayer form, the problem (7.28) of finding extremals may, however, be recast as that of minimizing the fixed-time optimization problem in Bolza form, that is,

$$\phi_0(t_0, \mathcal{X}(t_0)) = \mathrm{Tr}\left\{\mathcal{X}(t_0) z_0 z_0^T\right\} + \int_{t_0}^{t_f} \mathrm{Tr}\left\{\mathcal{X}(t) G_a(t) W G_a^T(t)\right\} \mathrm{d}t \tag{7.35}$$

subject to

$$\frac{\mathrm{d}}{\mathrm{d}\tau}\mathscr{X}(\tau) = -F_a^T(\tau)\mathscr{X}(\tau) - \mathscr{X}(\tau)F_a(\tau) - \mu_1 N(\tau)$$
$$-\sum_{r=2}^{k}\mu_r\sum_{s=1}^{r-1}\frac{2r!}{s!(r-s)!}\mathscr{H}_s(\tau)G_a(\tau)WG_a^T(\tau)\mathscr{H}_{r-s}(\tau), \quad (7.36)$$

wherein $\mathscr{X}(\tau) \triangleq \mu_1\mathscr{H}_1(\tau) + \cdots + \mu_k\mathscr{H}_k(\tau)$ and $\{\mathscr{H}_r(\tau)\}_{r=1}^k$ are satisfying the dynamical equations (7.22)–(7.24) with the initial-valued condition $\mathscr{X}(t_f) = \mu_1 N_f$ for all $\tau \in [t_0, t_f]$.

Furthermore, the transformation of problem (7.35) and (7.36) into the framework required by the matrix minimum principle [5] that makes it possible to apply Pontryagin's results directly to problems whose state variables are most conveniently regarded as matrices is complete if further changes of variables are introduced, that is, $t_f + t_0 - t = \tau$ and $\mathscr{X}(t_f + t_0 - t) = \mathscr{M}(t)$. Thus, the aggregate equation (7.36) with the initial-valued condition $\mathscr{M}(t_0) = \mu_1 N_f$ is rewritten:

$$\frac{\mathrm{d}}{\mathrm{d}t}\mathscr{M}(t) = F_a^T(t)\mathscr{M}(t) + \mathscr{M}(t)F_a(t) + \mu_1 N(t)$$
$$+\sum_{r=2}^{k}\mu_r\sum_{s=1}^{r-1}\frac{2r!}{s!(r-s)!}\mathscr{H}_s(t)G_a(t)WG_a^T(t)\mathscr{H}_{r-s}(t). \quad (7.37)$$

Now the aggregate matrix coefficients $F_a(t)$ and $N(t)$ for $t \in [t_0, t_f]$ of the composite dynamics (7.7) with trajectory sensitivity consideration are next partitioned to conform with the n-dimensional structure of Eq. (7.1) by means of $I_0^T \triangleq \begin{bmatrix} I\ 0 \end{bmatrix}$ and $I_1^T \triangleq \begin{bmatrix} 0\ I \end{bmatrix}$, whereby I is an $n \times n$ identity matrix and

$$F_a(t) = I_0(A(t,\zeta) + B(t,\zeta)K(t))I_0^T + I_1(A_\zeta(t,\zeta_0) + B_\zeta(t,\zeta_0)K(t))I_0^T$$
$$+ I_1(A(t,\zeta_0) + B(t,\zeta_0)K(t))I_1^T, \quad (7.38)$$
$$N(t) = I_0(Q(t) + K^T(t)R(t)K(t))I_0^T + I_1 S(t) I_1^T. \quad (7.39)$$

Assume that $\mathscr{K}_{t_f,\mathscr{H}_f,\mathscr{D}_f;\mu}$ is nonempty and convex in $\mathbb{R}^{m\times n}$. For all $(t,K) \in [t_0,t_f] \times \mathscr{K}_{t_f,\mathscr{H}_f,\mathscr{D}_f;\mu}$, the maps $h(\mathscr{M})$ and $q(t,\mathscr{M}(t,K))$ having the property of twice continuously differentiable, as defined from the risk-averse performance index (7.35)

$$\phi_0(K(\cdot)) = h(\mathscr{M}(t_f)) + \int_{t_0}^{t_f} q(t,\mathscr{M}(t,K(t)))\mathrm{d}t$$
$$= \mathrm{Tr}\left\{\mathscr{M}(t_f)z_0 z_0^T\right\} + \int_{t_0}^{t_f}\mathrm{Tr}\left\{\mathscr{M}(t)G_a(t)WG_a^T(t)\right\}\mathrm{d}t \quad (7.40)$$

are supposed to have all partial derivatives with respect to $\mathscr{M}$ up to order 2 being continuous in $(\mathscr{M}, K)$ with appropriate growths.

Moreover, any admissible feedback gain $K^* \in \mathscr{K}_{t_f,\mathscr{H}_f,\mathscr{D}_f;\mu}$ minimizing the risk-value aware performance index (7.40) is referred as an optimal strategies with risk aversion of the optimization problem (7.28). The corresponding state process $\mathscr{M}^*(\cdot)$ is called an optimal state process. Further denote $\mathscr{P}(t)$ by the co state matrix associated with $\mathscr{M}(t)$ for each $t \in [t_0, t_f]$. The scalar-valued Hamiltonian function for the optimization problem (7.37) and (7.40) is thus defined by

$$\mathscr{V}(t, \mathscr{M}, K) \triangleq \mathrm{Tr}\{\mathscr{M} G_a W G_a^T\} + \mathrm{Tr}\left\{\left[F_a^T \mathscr{M} + \mathscr{M} F_a + \mu_1 N + \sum_{r=2}^{k} \mu_r \sum_{s=1}^{r-1} \frac{2r!}{s!(r-s)!} \mathscr{H}_s G_a W G_a^T \mathscr{H}_{r-s}\right] \mathscr{P}^T(t)\right\}, \tag{7.41}$$

whereby in view of the expressions (7.38)–(7.39), the matrix variables $\mathscr{M}$, F_a, and N shall be considered as $\mathscr{M}(t,K)$, $F_a(t,K)$, and $N(t,K)$, respectively.

Using the matrix minimum principle [5], the set of first-order necessary conditions for K^* to be an extremizer is composed of

$$\frac{\mathrm{d}}{\mathrm{d}t}\mathscr{M}^*(t) = \left.\frac{\partial \mathscr{V}}{\partial \mathscr{P}}\right|_* = (F_a^*)^T(t)\mathscr{M}^*(t) + \mathscr{M}^*(t)F_a^*(t) + \mu_1 N^*(t) + \sum_{r=2}^{k} \mu_r \sum_{s=1}^{r-1} \frac{2r!}{s!(r-s)!} \mathscr{H}_s^*(t) G_a(t) W G_a^T(t) \mathscr{H}_{r-s}^*(t) \tag{7.42}$$

and

$$\frac{\mathrm{d}}{\mathrm{d}t}\mathscr{P}^*(t) = -\left.\frac{\partial \mathscr{V}}{\partial \mathscr{M}}\right|_* = -F_a^*(t)\mathscr{P}^*(t) - \mathscr{P}^*(t)(F_a^*)^T(t) - G_a(t) W G_a^T(t) \tag{7.43}$$

whereby the two-point boundary conditions $\mathscr{M}^*(t_0) = \mu_1 N_f$ and $\mathscr{P}^*(t_f) = z_0 z_0^T$.

Moreover, if K^* is a local extremum of Eq. (7.41), it implies

$$\mathscr{V}(t, \mathscr{M}^*(t), K) - \mathscr{V}(t, \mathscr{M}^*(t), K^*(t)) \geq 0 \tag{7.44}$$

for all $K \in \mathscr{K}_{t_f,\mathscr{H}_f,\mathscr{D}_f;\mu}$ and $t \in [t_0, t_f]$. That is,

$$\min_{K \in \mathscr{K}_{t_f,\mathscr{H}_f,\mathscr{D}_f;\mu}} \mathscr{V}(t, \mathscr{M}^*(t), K) = \mathscr{V}(t, \mathscr{M}^*(t), K^*(t)) = 0, \quad \forall t \in [t_0, t_f]. \tag{7.45}$$

Equivalently, it follows that

$$0 \equiv \left.\frac{\partial \mathscr{V}}{\partial K}\right|_* = -2\left[B^T(t,\zeta) I_0^T + B_\zeta^T(t,\zeta_0) I_1^T\right] \mathscr{M}^*(t)\mathscr{P}^*(t) I_0 - 2\mu_1 R(t) K I_0^T \mathscr{P}^*(t) I_0. \tag{7.46}$$

Furthermore, the second-order sufficient conditions that ensure the Hamiltonian functional (7.41) achieving its local minimum, require the following Hessian matrix to be positive definite; in particular,

$$\left.\frac{\partial^2 \mathscr{V}}{\partial K^2}\right|_* = 2\mu_1 R(t) \otimes I_0^T \mathscr{P}^*(t) I_0, \tag{7.47}$$

whereby the symbol $\otimes$ stands for the Kronecker matrix product operator.

By the matrix variation of constants formula [6], the matrix solutions of the cumulant-generating equations (7.22)–(7.23) and the co-state equation (7.43) can be rewritten in the integral forms, for each $\tau \in [t_0, t_f]$

$$\mathscr{H}_1^*(\tau) = \Phi^T(t_f,\tau) N_f \Phi(t_f,\tau) + \int_\tau^{t_f} \Phi(t_f,t) N^*(t) \Phi(t_f,t) dt, \tag{7.48}$$

$$\mathscr{H}_r^*(\tau) = \int_\tau^{t_f} \Phi(t_f,t) \sum_{s=1}^{r-1} \frac{2r!}{s!(r-s)!} \mathscr{H}_s^*(t) G_a(t) W G_a^T(t) \mathscr{H}_{r-s}^*(t) \Phi(t_f,t) dt, \tag{7.49}$$

$$\mathscr{P}^*(\tau) = \Phi^T(t_f,\tau) z_0 z_0^T \Phi(t_f,\tau) + \int_\tau^{t_f} \Phi(t_f,t) G_a(t) W G_a^T(t) \Phi(t_f,t) dt \tag{7.50}$$

provided that $\frac{d}{dt}\Phi(t,t_0) = F_a^*(t)\Phi(t,t_0)$ and $\Phi(t_0,t_0) = I$.

Interestingly enough, it can be verified that the following matrix inequalities, including $N_f \geq 0$, $N^*(t) > 0$, $z_0 z_0^T \geq 0$, $G_a(t) W G_a^T(t) > 0$ and

$$\sum_{s=1}^{r-1} \frac{2r!}{s!(r-s)!} \mathscr{H}_s^*(t) G_a(t) W G_a^T(t) \mathscr{H}_{r-s}^*(t) \geq 0$$

hold for all $t \in [t_0, t_f]$.

Therefore, it implies that $\{\mathscr{H}_r^*(t)\}_{r=1}^k$ and thus $\mathscr{M}^*(t)$, as well as $\mathscr{P}^*(t)$ with the integral forms (7.48)–(7.50) are positive definite on $[t_0, t_f]$. Subsequently, the following matrix inequality is shown to be valid

$$I_0^T \mathscr{P}^*(t) I_0 > 0, \qquad \forall t \in [t_0, t_f]. \tag{7.51}$$

In view of Eq. (7.51), it is concluded that the Hessian matrix (7.47) is indeed positive definite, in addition with the fact of $R(t)$ being positive definite for all $t \in [t_0, t_f]$. As the result, the local extremizer K^* formed by the first order necessary condition (7.46) becomes a local minimizer.

To this end, the result (7.42) is coupled by the forward-in-time and backward-in-time matrix-valued differential equations. Putting the corresponding state and co state equations together, the following optimality system with risk aversion and low sensitivity is summarized as follows:

Theorem 7.4.1 (Risk-Averse Control with Low Sensitivity). *Consider the class of stochastic linear systems described by Eqs. (7.1)–(7.3) wherein the pair (A,B) is uniformly stablizable. Then, the low-sensitivity feedback strategy $u(\cdot) = K(\cdot)x(\cdot)$ with risk aversion supported by the optimal feedback gain $K^*(\cdot)$ is given by*

$$K^*(t) = -R^{-1}(t)[B^T(t,\zeta)I_0^T \sum_{r=1}^{k} \hat{\mu}_r \mathscr{H}_r^*(t)\mathscr{P}^*(t_0+t_f-t)I_0 + B_\zeta^T(t,\zeta_0)I_1^T \sum_{r=1}^{k} \hat{\mu}_r \mathscr{H}_r^*(t)\mathscr{P}^*(t_0+t_f-t)I_0][I_0^T\mathscr{P}^*(t_0+t_f-t)I_0]^{-1}, \tag{7.52}$$

whereby the normalized parametric design of freedom $\hat{\mu}_r \triangleq \frac{\mu_r}{\mu_1}$ and the optimal state solutions $\{\mathscr{H}_r^*(\cdot)\}_{r=1}^k$ supporting performance robustness and risk-averse decisions are governed by the forward-in-time matrix-valued differential equations

$$\frac{d}{dt}\mathscr{H}_1^*(t) = (F_a^*)^T(t)\mathscr{H}_1^*(t) + \mathscr{H}_1^*(t)F_a^*(t) + N^*(t), \quad \mathscr{H}_1^*(t_0) = N_f, \tag{7.53}$$

$$\frac{d}{dt}\mathscr{H}_r^*(t) = (F_a^*)^T(t)\mathscr{H}_r^*(t) + \mathscr{H}_r^*(t)F_a^*(t) + \sum_{s=1}^{r-1} \frac{2r!}{s!(r-s)!}\mathscr{H}_s^*(t)G_a(t)WG_a^T(t)\mathscr{H}_{r-s}^*(t), \quad \mathscr{H}_r^*(t_0) = 0 \tag{7.54}$$

and the optimal co state solution $\mathscr{P}^*(\cdot)$ satisfies the backward-in-time matrix-valued differential equation with the terminal-value condition $\mathscr{P}^*(t_f) = z_0 z_0^T$:

$$\frac{d}{dt}\mathscr{P}^*(t) = -F_a^*(t)\mathscr{P}^*(t) - \mathscr{P}^*(t)(F_a^*)^T(t) - G_a(t)WG_a^T(t). \tag{7.55}$$

The results herein are certainly viewed as the generalization of those obtained from [3]. With respect to the subject of performance robustness, the previously developed work has fundamentally focused on the higher-order assessment of performance variations. Hence, the work therein [3] has not yet addressed the system trajectory sensitivity with respect to constant parameter variables.

7.5 Chapter Summary

The optimal risk-averse control strategy with low sensitivity has been obtained. A two-point boundary value problem involving matrix differential equations must be solved. Moreover, the states $\{\mathscr{H}_r(\cdot)^*\}_{r=1}^k$ and co state $\mathscr{P}^*(\cdot)$ play an important role in the determination of feedback strategies with risk aversion. It is important to note that the feedback gain $K^*(\cdot)$ depends on the mathematical statistics associated with performance uncertainty, in particular, mean, variance, and skewness. These statistics serve not only as feedback information for future risk-averse decisions but also as an influence mechanism for the low-sensitivity controller.

Finally, it is conjectured that an observer or a dynamical compensator could be used to implement a low-sensitivity control with risk aversion. The corresponding problem would then be cast into an output feedback form and thus will be the emerging subject of future research investigation.

References

1. D'Angelo, H., Moe, M.L., Hendricks, T.C.: Trajectory sensitivity of an optimal control systems. In: Proceedings of the 4th Allerton Conference on Circuit and Systems Theory, pp. 489–498 (1966)
2. Kahne, S.J.: Low-sensitivity design of optimal linear control systems. IEEE Trans. Aero. Electron. Syst. **4**(3), 374–379 (1968)
3. Pham, K.D., Sain, M.K., Liberty, S.R.: Cost cumulant control: state-feedback, finite-horizon paradigm with application to seismic protection. In: Miele, A. (ed.) Special Issue of J. Optim. Theor. Appl. **115**(3), 685–710 (2002), Kluwer Academic/Plenum Publishers, New York
4. Pham, K.D.: New risk-averse control paradigm for stochastic two-time-scale systems and performance robustness. In: Miele, A. (ed.) J. Optim. Theor. Appl. **146**(2), 511–537 (2010)
5. Athans, M.: The matrix minimum principle. IEEE Trans. Inform. Contr. **11**, 592–606 (1968)
6. Brockett, R.W.: Finite Dimensional Linear Systems. Wiley, New York (1970)

Chapter 8
Output-Feedback Control for Stochastic Systems with Low Sensitivity

Abstract The chapter treats the problem of controlling stochastic linear systems with quadratic criteria, including sensitivity variables, when noisy measurements are available. It is proved that the low sensitivity control strategy with risk aversion can be realized by the cascade of (1) the conditional mean estimate of the current state using a Kalman-like estimator and (2) optimally feedback, which is effectively supported by the mathematical statistics of performance uncertainty as if the conditional mean state estimate was the true state of the system. In other words, the certainty equivalence principle still holds for this statistical optimal control problem.

8.1 Introduction

Looking critically at the literature in parameter sensitivity reductions in linear regulators for the last 40 years, the traditional approach to trajectory and performance index sensitivity reductions seems to be the winner because its main objective is to achieve a tradeoff between optimality in the nominal performance and sensitivity to small parameter variations. It is appropriate to recall some of pioneering results, summarized in [1, 2] as representative examples, that were published in the early 1960s. In these early works, the importance including of a quadratic trajectory sensitivity term in the integrand of the performance index was advocated. Unfortunately, the theory only handles rather simple performance criteria wherein the conventional measure of performance variations is largely centered on statistical average or mean and thus, may not be accurate enough to describe complex behavior of closed-loop performance distributions of today's real world problems.

With this perspective in mind, the research investigation herein considers a linear-quadratic problem class of controlling stochastic linear systems with output feedback and multi-level performance robustness under the quadratic criteria, including the system state sensitivity variables with respect to some constant parameters. The innovative technical approach as proposed in this research investigation,

K.D. Pham, *Linear-Quadratic Controls in Risk-Averse Decision Making*,
SpringerBriefs in Optimization, DOI 10.1007/978-1-4614-5079-5_8,

to solutions for the analysis and decision problems [3], within which any evaluation scheme based on just the expected outcome and outcome variance would necessarily imply indifference between some courses of action, rests upon: (1) acquisition and utilization of insights about performance uncertainty, regarding whether the control designers are risk averse or risk prone, and (2) best courses of action to ensure performance robustness and reliability, provided that the control designers be subscribed to certain attitudes.

The structure of this chapter is as follows. Section 8.2 addresses the control problem and the notion of admissible control. In addition, the development of all the mathematical statistics for performance robustness is also summarized. Sections 8.3 and 8.4 are directed to the attention of problem statements and the solution method via an adapted dynamic programming. The chapter ends with Sect. 8.5, whereby some conclusions are further included.

8.2 Linking Performance-Measure Statistics and Risk Perceptions

In this section, some preliminaries are in order. Some spaces of random variables and stochastic processes are introduced: $L^2_{\mathscr{F}_t}(\Omega;\mathbb{R}^n) \triangleq \{\wp : \Omega \mapsto \mathbb{R}^n$ such that $\wp$ is $\mathscr{F}_t$-measurable and $E\{\|\wp\|^2\} < \infty\}$. In addition, $L^2_{\mathscr{F}}([t_0,t_f];\mathbb{R}^k) \triangleq \{\hbar : [t_0,t_f] \times \Omega \mapsto \mathbb{R}^k$ such that $\hbar(\cdot)$ is $\mathbb{F} = \{\mathscr{F}_t\}_{t \ge t_0 \ge 0}$-adapted and $E\{\int_{t_0}^{t_f} \|\hbar(t)\|^2 \mathrm{d}t\} < \infty\}$.

Now, as for the setting, some classes of problems with incomplete information shall be investigated. Particularly, the stochastic system being controlled on a finite horizon $[t_0,t_f]$ is linear and depends on the constant parameter variable ζ with the known initial condition $x(t_0) = x_0$

$$\mathrm{d}x(t) = (A(t,\zeta)x(t) + B(t,\zeta)u(t))\mathrm{d}t + G(t)\mathrm{d}w(t) \tag{8.1}$$

to which, the control design has available measurements

$$\mathrm{d}y(t) = C(t)x(t)\mathrm{d}t + \mathrm{d}v(t). \tag{8.2}$$

Here $x(t) \equiv x(t,\omega)$ is the controlled state process valued in $\mathbb{R}^n$; $u(t) \equiv u(t,\omega)$ is the control process valued in an admissible set $U \subseteq L^2_{\mathscr{F}}([t_0,t_f];\mathbb{R}^m)$; and $y(t) \equiv y(t,\omega)$ is the measured output of the manipulated process $x(t)$.

In the problem description, as described by Eqs. (8.1) and (8.2), the system coefficients $A(t,\zeta) \equiv A(t,\zeta,\omega) : [t_0,t_f] \times \Omega \mapsto \mathbb{R}^{n\times n}$ and $B(t,\zeta) \equiv B(t,\zeta,\omega) : [t_0,t_f] \times \Omega \mapsto \mathbb{R}^{n\times m}$ are the continuous-time matrix functions of the time and constant parameter variables t and ζ, respectively. The disturbance distribution coefficient $G(t) \equiv G(t,\omega) : [t_0,t_f] \times \Omega \mapsto \mathbb{R}^{n\times p}$ and the measurement coefficient $C(t) \equiv C(t,\omega) : [t_0,t_f] \times \Omega \mapsto \mathbb{R}^{r\times n}$ are continuous-time matrix functions independent of ζ. The random process and measurement disturbances $w(t) \equiv w(t,\omega)$ and

$v(t) \equiv v(t,\omega)$ are assumed to be the correlated stationary Wiener processes with the correlations of independent increments for all $\tau_1, \tau_2 \in [t_0, t_f]$

$$E\left\{[w(\tau_1) - w(\tau_2)][w(\tau_1) - w(\tau_2)]^T\right\} = W|\tau_1 - \tau_2|,$$

$$E\left\{[v(\tau_1) - v(\tau_2)][v(\tau_1) - v(\tau_2)]^T\right\} = V|\tau_1 - \tau_2|,$$

$$E\left\{[w(\tau_1) - w(\tau_2)][v(\tau_1) - v(\tau_2)]^T\right\} = N|\tau_1 - \tau_2|$$

whose a priori second-order statistics $W > 0$, $V > 0$ and $N > 0$ are also known.

Notice that the state $x(t)$ is considered as the function of the constant parameter variable ζ, i.e., $x(t) \equiv \pi(t, x_0, \zeta)$. The traditional approach to sensitivity analysis for the system trajectory has been to define a sensitivity variable by

$$\sigma(t) \triangleq \frac{\partial}{\partial \zeta}\pi(t, x_0, \zeta)\bigg|_{\zeta=\zeta_0}. \tag{8.3}$$

Associated with each feasible $u \in U$ is a path-wise finite-horizon integral-quadratic-form (IQF) performance measure with the generalized chi-squared random distribution $J : U \mapsto \mathbb{R}^+$, for which the control designer attempts to control the system with multiple degrees of performance robustness

$$J(u(\cdot)) = x^T(t_f)Q_f x(t_f) + \int_{t_0}^{t_f} [x^T(\tau)Q(\tau)x(\tau)$$
$$+ u^T(\tau)R(\tau)u(\tau) + \sigma^T(\tau)S(\tau)\sigma(\tau)]d\tau, \tag{8.4}$$

whereby the terminal penalty weighting $Q_f \in \mathbb{R}^{n\times n}$, the state weighting $Q(t) \equiv Q(t,\omega) : [t_0,t_f] \times \Omega \mapsto \mathbb{R}^{n\times n}$, the trajectory sensitivity weighting $S(t) \equiv S(t,\omega) : [t_0,t_f] \times \Omega \mapsto \mathbb{R}^{n\times n}$ and control weighting $R(t) \equiv R(t,\omega) : [t_0,t_f] \times \Omega \mapsto \mathbb{R}^{m\times m}$ are continuous-time matrix functions with the properties of symmetry and positive semidefiniteness. In addition, $R(t)$ is invertible.

Next the notion of admissible controls is discussed. In the case of incomplete information, an admissible control must be of the form, for some $\eth(\cdot,\cdot)$

$$u(t) = \eth(t, y(\tau)), \quad \tau \in [t_0, t]. \tag{8.5}$$

In general, the conditional probability density $p(x(t)|\mathscr{F}_t)$ which is the probability density of $x(t)$ conditioned on $\mathscr{F}_t$ (i.e., induced by the observation $\{y(\tau) : \tau \in [t_0,t]\}$) represents the sufficient statistics for describing the conditional stochastic effects of future control action. Under the Gaussian assumption, the conditional probability density $p(x(t)|\mathscr{F}_t)$ is parameterized by its conditional mean $\hat{x}(t) \triangleq E\{x(t)|\mathscr{F}_t\}$ and covariance $\Sigma(t) \triangleq E\{[x(t) - \hat{x}(t)][x(t) - \hat{x}(t)]^T|\mathscr{F}_t\}$. With respect to the linear-Gaussian conditions, the covariance is independent of control and observation. Therefore, to look for an optimal control law of the form Eq. (8.5), it is only required that

$$u(t) = \gamma(t, \hat{x}(t)), \quad t \in [t_0, t_f].$$

As shown in [6], the search for optimal control solutions to the statistical optimal control problem may be consistently and productively restricted to linear time-varying feedback laws generated from the accessible state $\hat{x}(t)$ by

$$u(t) = K(t)\hat{x}(t), \quad t \in [t_0, t_f] \tag{8.6}$$

with $K \in C([t_0,t_f];\mathbb{R}^{m\times n})$ an admissible gain whose further defining properties will be stated shortly.

Then, for the admissible feedback gain $K(\cdot)$ and pair (t_0, x_0), it gives a sufficient condition for the existence of $x(t)$ in Eq. (8.1). In view of the feedback law (8.6), the controlled system (8.1) with the initial condition $x(t_0) = x_0$ is rewritten as follows

$$\mathrm{d}x(t) = (A(t,\zeta) + B(t,\zeta)K(t))\hat{x}(t)\mathrm{d}t + G(t)\mathrm{d}w(t). \tag{8.7}$$

In addition, the variation of the parameter ζ from the nominal value ζ_0 was assumed to be small, such that the Taylor's expansion of the state and control functions can be approximated by retaining only the first two terms. Differentiating the controlled process (8.7) with respect to ζ and evaluating at $\zeta = \zeta_0$ gives the initial-value condition $\sigma(t_0) = 0$ and

$$\mathrm{d}\sigma(t) = (A_\zeta(t,\zeta_0) + B_\zeta(t,\zeta_0)K(t))\hat{x}(t)\mathrm{d}t + (A(t,\zeta_0) + B(t,\zeta_0)K(t))\hat{\sigma}(t)\mathrm{d}t, \tag{8.8}$$

whereby $A_\zeta \triangleq \frac{\partial A}{\partial \zeta}\Big|_{\zeta=\zeta_0}$, $B_\zeta \triangleq \frac{\partial B}{\partial \zeta}\Big|_{\zeta=\zeta_0}$ and $\hat{\sigma} \triangleq \frac{\partial \hat{x}}{\partial \zeta}\Big|_{\zeta=\zeta_0}$.

Notice that the linear-Gaussian assumptions imply that the estimate-state mean $\hat{x}(t)$ and estimate-error covariance $\Sigma(t)$ are generated by the Kalman-like estimator with $\hat{x}(t_0) = x_0$

$$\mathrm{d}\hat{x}(t) = (A(t,\zeta) + B(t,\zeta)K(t))\hat{x}(t)\mathrm{d}t + L(t)(dy(t) - C(t)\hat{x}(t)\mathrm{d}t) \tag{8.9}$$

and the Kalman-filtering gain $L(\cdot)$ is given by

$$L(t) = [\Sigma(t)C^T(t) + G(t)N]V^{-1}, \tag{8.10}$$

$$\begin{aligned}\frac{\mathrm{d}}{\mathrm{d}t}\Sigma(t) &= A(t,\zeta)\Sigma(t) + \Sigma(t)A^T(t,\zeta) + G(t)WG^T(t)\\ &\quad - [\Sigma(t)C^T(t) + G(t)N]V^{-1}[C(t)\Sigma(t) + N^TG^T(t)], \quad \Sigma(t_0) = 0.\end{aligned} \tag{8.11}$$

Taking the partial derivative of Eq. (8.9) with respect to ζ yields

$$\begin{aligned}\mathrm{d}\hat{\sigma}(t) &= (A_\zeta(t,\zeta_0) + B_\zeta(t,\zeta_0)K(t))\hat{x}(t)\mathrm{d}t + (A(t,\zeta_0) + B(t,\zeta_0)K(t))\hat{\sigma}(t)\mathrm{d}t\\ &\quad + L_\zeta(t)C(t)\tilde{x}(t)\mathrm{d}t + L(t)C(t)\tilde{\sigma}(t)\mathrm{d}t + L_\zeta(t)dv(t), \quad \hat{\sigma}(t_0) = 0\end{aligned} \tag{8.12}$$

where the estimate errors $\tilde{x}(t) \triangleq x(t) - \hat{x}(t)$, $\tilde{\sigma}(t) \triangleq \left.\frac{\partial \tilde{x}}{\partial \zeta}\right|_{\zeta=\zeta_0}$ and $L_\zeta \triangleq \left.\frac{\partial L}{\partial \zeta}\right|_{\zeta=\zeta_0}$.

Using the definitions for error estimates $\tilde{x}(t)$ and $\tilde{\sigma}(t)$, it is shown that

$$d\tilde{x}(t) = (A(t,\zeta) - L(t)C(t))\tilde{x}(t)dt + G(t)dw(t) - L(t)dv(t), \quad \tilde{x}(t_0) = 0 \quad (8.13)$$

and

$$\begin{aligned} d\tilde{\sigma}(t) &= (A_\zeta(t,\zeta_0) - L_\zeta(t)C(t))\tilde{x}(t)dt - L_\zeta(t)dv(t) \\ &\quad + (A(t,\zeta_0) - L(t)C(t))\tilde{\sigma}(t)dt, \quad \tilde{\sigma}(t_0) = 0, \end{aligned} \quad (8.14)$$

whereby $L_\zeta(t) = \Sigma_\zeta(t)C^T(t)V^{-1}$ and $\Sigma_\zeta \triangleq \left.\frac{\partial \Sigma}{\partial \zeta}\right|_{\zeta=\zeta_0}$ is satisfying

$$\begin{aligned} \frac{d}{dt}\Sigma_\zeta(t) &= A_\zeta(t,\zeta_0)\Sigma(t) + A(t,\zeta_0)\Sigma_\zeta(t) + \Sigma_\zeta(t)A^T(t,\zeta_0) \\ &\quad + \Sigma(t)A_\zeta^T(t,\zeta_0) - \Sigma_\zeta(t)C^T(t)V^{-1}[C(t)\Sigma(t) + N^T G^T(t)] \\ &\quad - [\Sigma(t)C(t) + G(t)N]V^{-1}C(t)\Sigma_\zeta(t), \quad \Sigma_\zeta(t_0) = 0. \end{aligned} \quad (8.15)$$

To progress towards the analysis of performance reliability, the aggregate dynamics is, therefore, governed by the controlled stochastic differential equation

$$dz(t) = F_a(t)z(t)dt + G_a(t)dw_a(t), \quad z(t_0) = z_0, \quad (8.16)$$

whereby, for each $t \in [t_0, t_f]$, the augmented state variables $z^T = \begin{bmatrix} \hat{x}^T & \tilde{x}^T & \hat{\sigma}^T & \tilde{\sigma}^T \end{bmatrix}$, the aggregate process noises $w_a^T = \begin{bmatrix} w^T & v^T \end{bmatrix}$ with the correlation of independent increments defined as

$$E\left\{[w_a(\tau_1) - w_a(\tau_2)][w_a(\tau_1) - w_a(\tau_2)]^T\right\} = W_a|\tau_1 - \tau_2|, \forall \tau_1, \tau_2 \in [t_0, t_f]$$

and the augmented system coefficients are given by

$$F_a = \begin{bmatrix} F_a^{11} & F_a^{12} & F_a^{13} & F_a^{14} \\ F_a^{21} & F_a^{22} & F_a^{23} & F_a^{24} \\ F_a^{31} & F_a^{32} & F_a^{33} & F_a^{34} \\ F_a^{41} & F_a^{42} & F_a^{43} & F_a^{44} \end{bmatrix}, \; G_a = \begin{bmatrix} 0 & L \\ G & -L \\ 0 & L_\zeta \\ 0 & -L_\zeta \end{bmatrix}, \; z_0 = \begin{bmatrix} x_0 \\ 0 \\ 0 \\ 0 \end{bmatrix}, \; W_a = \begin{bmatrix} W & N \\ N^T & V \end{bmatrix}$$

provided that $F_a^{11}(t) = A(t,\zeta) + B(t,\zeta)K(t)$, $F_a^{12}(t) = L(t)C(t)$, $F_a^{13}(t) = F_a^{14}(t) = 0$; $F_a^{21}(t) = F_a^{23}(t) = F_a^{24}(t) = 0$, $F_a^{22}(t) = A(t,\zeta) - L(t)C(t)$; $F_a^{31}(t) = A_\zeta(t,\zeta_0) + B_\zeta(t,\zeta_0)K(t)$, $F_a^{32}(t) = L_\zeta(t)C(t)$, $F_a^{33}(t) = A(t,\zeta_0) + B(t,\zeta_0)K(t)$, $F_a^{34}(t) = L(t)C(t)$; and $F_a^{41}(t) = F_a^{43}(t) = 0$, $F_a^{42}(t) = A_\zeta(t,\zeta_0) - L_\zeta(t)C(t)$, and $F_a^{44}(t) = A(t,\zeta_0) - L(t)C(t)$ for all $t \in [t_0, t_f]$ and some ζ and ζ_0.

In addition, the performance cost (8.4) is rewritten as follows

$$J(K(\cdot)) = z^T(t_f)N_f z(t_f) + \int_{t_0}^{t_f} z^T(\tau)N_a(\tau)z(\tau)\mathrm{d}\tau, \tag{8.17}$$

whereby the continuous-time weightings are given by

$$N_f = \begin{bmatrix} Q_f & Q_f & 0 & 0 \\ Q_f & Q_f & 0 & 0 \\ 0 & 0 & 0 & 0 \\ 0 & 0 & 0 & 0 \end{bmatrix}, \quad N_a(t) = \begin{bmatrix} Q(t)+K^T(t)R(t)K(t) & Q(t) & 0 & 0 \\ Q(t) & Q(t) & 0 & 0 \\ 0 & 0 & S(t) & S(t) \\ 0 & 0 & S(t) & S(t) \end{bmatrix}.$$

Given the lack of analysis of performance risk and stochastic preferences, much of the discussion that follows will concern the case involving performance riskiness from the standpoint of higher-order characteristics pertaining to performance sampling distributions. Notice that the performance measure (8.17) is clearly a random variable of the generalized chi-squared type. Hence, the degrees of performance uncertainty of the random cost (8.17) must be assessed by modeling and management of cost cumulants (also known as semi-invariants) associated with the random cost (8.17). Precisely stated, these cost cumulants can be generated by the use of the Maclaurin series expansion of the cumulant-generating function of the random variab (8.17). The technical merits of this analysis will provide an effective and accurate capability for forecasting all the higher-order cumulants associated with performance uncertainty. As expected, via these higher-order performance-measure statistics, the controller design with performance risk aversion is thus capable of reshaping the cumulative probability distribution of future closed-loop performance.

Theorem 8.2.1 (Cumulant-Generating Function). *Let $z(\cdot)$ be a state variable of the stochastic dynamics concerning sensitivity (8.16) with initial values $z(\tau) \equiv z_\tau$ and $\tau \in [t_0, t_f]$. Further let the moment-generating function be denoted by*

$$\varphi(\tau, z_\tau; \theta) = \rho(\tau, \theta) \exp\{z_\tau^T \Upsilon(\tau, \theta) z_\tau\}, \tag{8.18}$$

$$\upsilon(\tau, \theta) = \ln\{\rho(\tau, \theta)\}, \qquad \theta \in \mathbb{R}^+. \tag{8.19}$$

Then, the cumulant-generating function has the form of quadratic affine

$$\psi(\tau, z_\tau; \theta) = z_\tau^T \Upsilon(\tau, \theta) z_\tau + \upsilon(\tau, \theta), \tag{8.20}$$

where the scalar solution $\upsilon(\tau, \theta)$ solves the backward-in-time differential equation with the terminal-value condition $\upsilon(t_f, \theta) = 0$

$$\frac{\mathrm{d}}{\mathrm{d}\tau}\upsilon(\tau, \theta) = -\mathrm{Tr}\{\Upsilon(\tau, \theta) G_a(\tau) W_a G_a^T(\tau)\} \tag{8.21}$$

and the matrix solution $\Upsilon(\tau,\theta)$ satisfies the backward-in-time differential equation with the terminal-value condition $\Upsilon(t_f,\theta) = \theta N_f$

$$\frac{\mathrm{d}}{\mathrm{d}\tau}\Upsilon(\tau,\theta) = -F_a^T(\tau)\Upsilon(\tau,\theta) - \Upsilon(\tau,\theta)F_a(\tau) - 2\Upsilon(\tau,\theta)G_a(\tau)W_aG_a^T(\tau)\Upsilon(\tau,\theta) - \theta N_a(\tau). \quad (8.22)$$

Meanwhile, the scalar solution $\rho(\tau,\theta)$ satisfies the backward-in-time differential equation with the terminal-value condition $\rho\left(t_f,\theta\right) = 1$

$$\frac{\mathrm{d}}{\mathrm{d}\tau}\rho(\tau,\theta) = -\rho(\tau,\theta)\,\mathrm{Tr}\left\{\Upsilon(\tau,\theta)G_a(\tau)W_aG_a^T(\tau)\right\}. \quad (8.23)$$

Proof. Needless to stress the technical developments of the proof herein are well established in Chap. 7 and Sect. 7.2. □

Subsequently, higher-order statistics associated with the random cost (8.17) can be utilized to quantify the stochastic nature of performance uncertainty beyond first and/or second-order statistics for closed-loop performance reliability without full-blown Monte Carlos performance dispersion analysis that requires a much larger computational cost. The determination of these mathematical statistics underpinning complex behavior of the random cost (8.17) is possible because of a Maclaurin series expansion of the cumulant-generating function (8.20), e.g.,

$$\psi(\tau,z_\tau;\theta) = \sum_{r=1}^{\infty}\left.\frac{\partial^{(r)}}{\partial\theta^{(r)}}\psi(\tau,z_\tau;\theta)\right|_{\theta=0}\frac{\theta^r}{r!} \quad (8.24)$$

in which all $\kappa_r \triangleq \left.\frac{\partial^{(r)}}{\partial\theta^{(r)}}\psi(\tau,z_\tau,\theta)\right|_{\theta=0}$ are called rth-order performance-measure statistics. Moreover, the series expansion coefficients are computed by using the cumulant-generating function (8.20)

$$\left.\frac{\partial^{(r)}}{\partial\theta^{(r)}}\psi(\tau,z_\tau;\theta)\right|_{\theta=0} = z_\tau^T\left.\frac{\partial^{(r)}}{\partial\theta^{(r)}}\Upsilon(\tau,\theta)\right|_{\theta=0}z_\tau + \left.\frac{\partial^{(r)}}{\partial\theta^{(r)}}\upsilon(\tau,\theta)\right|_{\theta=0}. \quad (8.25)$$

In view of the definition (8.24), the rth performance-measure statistic becomes

$$\kappa_r = z_\tau^T\left.\frac{\partial^{(r)}}{\partial\theta^{(r)}}\Upsilon(\tau,\theta)\right|_{\theta=0}z_\tau + \left.\frac{\partial^{(r)}}{\partial\theta^{(r)}}\upsilon(\tau,\theta)\right|_{\theta-0} \quad (8.26)$$

for any finite $1 \le r < \infty$. For notational convenience, it is necessary to take a step further along the way towards following change of notations

$$H_r(\tau) \triangleq \left.\frac{\partial^{(r)}\Upsilon(\tau,\theta)}{\partial\theta^{(r)}}\right|_{\theta=0}; \qquad D_r(\tau) \triangleq \left.\frac{\partial^{(r)}\upsilon(\tau,\theta)}{\partial\theta^{(r)}}\right|_{\theta=0}. \quad (8.27)$$

The emerging problem of statistical optimal control design being formulated then simply requires a new model concept for performance benefits and risks, which essentially consists of trade-offs between performance values and risks for reliable decisions. It therefore underlies a risk aversion notion in performance assessment and feedback decision design under uncertainty.

Theorem 8.2.2 (Performance-Measure Statistics). *Let the linear-quadratic stochastic system be described by Eqs. (8.16) and (8.17) in which the pairs (A,B) and (A,C) are uniformly stabilizable and detectable, respectively. For $k \in \mathbb{Z}^+$ fixed, the kth cumulant of performance measure (8.17) is*

$$\kappa_k = z_0^T H_k(t_0) z_0 + D_k(t_0), \tag{8.28}$$

where the supporting variables $\{H_r(\tau)\}_{r=1}^k$ and $\{D_r(\tau)\}_{r=1}^k$ evaluated at $\tau = t_0$ satisfy the differential equations (with the dependence of $H_r(\tau)$ and $D_r(\tau)$ upon $K(\tau)$ suppressed)

$$\frac{d}{d\tau} H_1(\tau) = -F_a^T(\tau) H_1(\tau) - H_1(\tau) F_a(\tau) - N_a(\tau), \tag{8.29}$$

$$\begin{aligned} \frac{d}{d\tau} H_r(\tau) &= -F_a^T(\tau) H_r(\tau) - H_r(\tau) F_a(\tau) \\ &\quad - \sum_{s=1}^{r-1} \frac{2r!}{s!(r-s)!} H_s(\tau) G_a(\tau) W_a G_a^T(\tau) H_{r-s}(\tau), \quad 2 \le r \le k, \end{aligned} \tag{8.30}$$

$$\frac{d}{d\tau} D_r(\tau) = -\mathrm{Tr}\left\{ H_r(\tau) G_a(\tau) W_a G_a^T(\tau) \right\}, \quad 1 \le r \le k, \tag{8.31}$$

whereby the terminal-value conditions $H_1(t_f) = N_f$, $H_r(t_f) = 0$ for $2 \le r \le k$ and $D_r(t_f) = 0$ for $1 \le r \le k$.

Proof. The proof follows by parallel development as illustrated in Chap. 7 and Sect. 7.2. □

Theorem 8.2.3 (Effect of Terminal Time on the Monotonicity of Solutions). *Fix $k \in \mathbb{Z}^+$ and denote $H_r(\alpha, t_f) \in \mathscr{C}^1([t_0, t_f]; \mathbb{R}^{n \times n})$ by the solutions to Eqs. (8.29)–(8.30)*

$$\frac{d}{d\tau} H_1(\tau, t_f) = -F_a^T(\tau) H_1(\tau, t_f) - H_1(\tau, t_f) F_a(\tau) - N_a(\tau), \quad H_1(t_f, t_f) = N_f,$$

$$\begin{aligned} \frac{d}{d\tau} H_r(\tau, t_f) &= -F_a^T(\tau) H_r(\tau, t_f) - H_r(\tau, t_f) F_a(\tau) \\ &\quad - \sum_{s=1}^{r-1} H_s(\tau, t_f) G_a(\tau) W_a G_a^T(\tau) H_{r-s}(\tau, t_f), \qquad H_r(t_f, t_f) = 0. \end{aligned}$$

Suppose that the system (8.16) is exponentially stable and the terminal-value condition $N_f = 0$. *Then, the solutions* $\{H_r(\tau,t_f)\}_{r=1}^k$ *and* $\{H_r(\tau,t_f+\Delta t_f)\}_{r=1}^k$ *of Eqs. (8.29)–(8.30) satisfy the inequality*

$$H_r(\tau,t_f+\Delta t_f) \geq H_r(\tau,t_f),$$

for some $\Delta t_f > 0$ *and* $t_0 \leq \tau \leq t_f$. *In other words, the solutions* $H_r(\tau,t_f)$ *are monotone increasing as* $t_f \to \infty$.

Proof. It is easy to verify that the solutions of Eqs. (8.29)–(8.30) are of the integral form

$$H_1(\tau,t_f) = \Phi_a^T(t_f,\tau)N_f\Phi_a(t_f,\tau) + \int_\tau^{t_f} \Phi_a^T(\tau,\tau)N_a(\tau)\Phi_a(\tau,\tau)\mathrm{d}\tau,$$

$$H_r(\tau,t_f) = \int_\tau^{t_f} \Phi_a^T(\tau,\tau)\sum_{s=1}^{r-1} H_s(\tau,t_f)G_a(\tau)W_aG_a^T(\tau)H_{r-s}(\tau,t_f)\Phi_a(\tau,\tau)\mathrm{d}\tau$$

in which $\Phi_a(\cdot,\tau)$ is the state transition matrix of $F_a(\tau)$ and $\tau \in [t_0,t_f]$ is fixed.

For any given $\Delta t_f > 0$ and by the assumption of $N_f = 0$, the solutions are written as follows:

$$H_1(\tau,t_f+\Delta t_f) = \int_\tau^{t_f+\Delta t_f} \Phi_a^T(\tau,\tau)N_a(\tau)\Phi_a(\tau,\tau)\mathrm{d}\tau,$$

$$H_r(\tau,t_f+\Delta t_f) = \int_\tau^{t_f+\Delta t_f} \Phi_a^T(\tau,\tau)\sum_{s=1}^{r-1} H_s(\tau,t_f+\Delta t_f)$$
$$\times G_a(\tau)W_aG_a^T(\tau)H_{r-s}(\tau,t_f+\Delta t_f)\Phi_a(\tau,\tau)\mathrm{d}\tau, \quad 2 \leq r \leq k.$$

Or, equivalently

$$H_1(\tau,t_f+\Delta t_f) = \int_\tau^{t_f} \Phi_a^T(\tau,\tau)N_a(\tau)\Phi_a(\tau,\tau)\mathrm{d}\tau + \int_{t_f}^{t_f+\Delta t_f} \Phi_a^T(\tau,\tau)N_a(\tau)\Phi_a(\tau,\tau)\mathrm{d}\tau,$$

$$H(\tau,t_f+\Delta t_f,i) = \int_\tau^{t_f} \Phi_a^T(\tau,\tau)\sum_{s=1}^{r-1} H(\tau,t_f,s)G_a(\tau)W_aG_a^T(\tau)H_{r-s}(\tau,t_f)\Phi_a(\tau,\tau)\mathrm{d}\tau$$
$$+\int_{t_f}^{t_f+\Delta t_f} \Phi_a^T(\tau,\tau)\sum_{s=1}^{r-1} H_s(\tau,t_f+\Delta t_f)$$
$$\times G_a(\tau)W_aG_a^T(\tau)H_{r-s}(\tau,t_f+\Delta t_f)\Phi_a(\tau,\tau)\mathrm{d}\tau, \quad 2 \leq r \leq k.$$

By the forms of solutions and the semi-group property of the state transition matrix

$$\Phi_a(\tau,\tau) = \Phi_a(\tau,t_f)\Phi_a(t_f,\tau),$$

we rewrite the solutions with the terminal time t_f plus increment Δt_f as follows

$$H_1(\tau,t_f+\Delta t_f) = H_1(\tau,t_f) + \Phi_a^T(t_f,\tau)H_1(t_f,t_f+\Delta t_f)\Phi_a(t_f,\tau),$$
$$H_r(\tau,t_f+\Delta t_f) = H_r(\tau,t_f) + \Phi_a^T(t_f,\tau)H_r(t_f,t_f+\Delta t_f)\Phi_a(t_f,\tau), \quad 2 \le i \le k.$$

Then forming the difference between the matrix solutions yields

$$H_1(\tau,t_f+\Delta t_f) - H_1(\tau,t_f) = \Phi_a^T(t_f,\tau)H_1(t_f,t_f+\Delta t_f)\Phi_a(t_f,\tau),$$
$$H_r(\tau,t_f+\Delta t_f) - H_r(\tau,t_f) = \Phi_a^T(t_f,\tau)H_r(t_f,t_f+\Delta t_f)\Phi_a(t_f,\tau), \quad 2 \le r \le k.$$

Since $H_1(t_f,t_f+\Delta t_f)$ and $H_r(t_f,t_f+\Delta t_f)$ are positive semidefinite, it follows that

$$H_1(\tau,t_f+\Delta t_f) \ge H_1(\tau,t_f),$$
$$H_r(\tau,t_f+\Delta t_f) \ge H_r(\tau,t_f), \quad 2 \le r \le k$$

for any given $\Delta t_f > 0$ and t_f arbitrary large. Hence, the proof is now completed. □

Even if $N_f \neq 0$, the additional term which would have appeared in the matrix difference, $H_1(\tau,t_f+\Delta t_f) - H_1(\tau,t_f)$

$$\Phi_a^T(t_f+\Delta t_f,\tau)N_f\Phi_a(t_f+\Delta t_f,\tau) - \Phi_a^T(t_f,\tau)N_f\Phi_a(t_f,\tau)$$

can be reexpressed as

$$\Phi_a^T(t_f,\tau)\left[\Phi_a^T(t_f+\Delta t_f,t_f)N_f\Phi_a(t_f+\Delta t_f,t_f) - N_f\right]\Phi_a(t_f,\tau).$$

Therefore, it follows that as t_f gets large,

$$\Phi_a^T(t_f,\tau)\left[\Phi_a^T(t_f+\Delta t_f,t_f)N_f\Phi_a(t_f+\Delta t_f,t_f) - N_f\right]\Phi_a(t_f,\tau)$$

is vanished exponentially. Why? Because as t_f tends to infinity,

$$\Phi_a(t_f+\Delta t_f,t_f) \to I,$$
$$||\Phi_a(t_f,\tau)|| \to 0,$$

which then makes

$$\Phi_a^T(t_f,\tau)\left[\Phi_a^T(t_f+\Delta t_f,t_f)N_f\Phi_a(t_f+\Delta t_f,t_f) - N_f\right]\Phi_a(t_f,\tau) \to 0 \text{ exponentially.}$$

Hence, the terminal penalty weighting matrix N_f has no significant role in the infinite-horizon problem. Without loss of generality, N_f can thus assume zero.

Next, the solutions to the cumulant-generating equations are bounded above under the assumption of exponential stability.

Theorem 8.2.4 (Upper Bounds on Matrix Solutions). *Let $k \in \mathbb{Z}^+$ and the system (8.16) be exponentially stable. Then, Eqs. (8.29)–(8.30) admit solutions $\{H_r(\cdot)\}_{r=1}^k$ whose matrix norms are bounded above, e.g.*

$$||H_r(\tau)|| \le \Gamma_r, \qquad 1 \le r \le k, \tag{8.32}$$

where the upper bounds Γ_r are given by

$$\Gamma_1 = \overline{N}_a M_2,$$

$$\Gamma_r = \sum_{s=1}^{r-1} \Gamma_s \Gamma_{r-s} \overline{G}_a^2 ||W_a|| M_2, \quad 2 \le r \le k$$

and

$$\overline{N}_a \triangleq \max_{\tau \in [t_0, t_f]} ||N_a(\tau)||, \qquad \lim_{t_f \to \infty} \int_\tau^{t_f} ||\Phi_a(\sigma,\tau)||^2 \, d\sigma \le M_2,$$

$$\overline{G}_a \triangleq \max_{\tau \in [t_0, t_f]} ||G_a(\tau)||.$$

Proof. For all $\tau \in [t_0, t_f]$, the integral solution form of $H_1(\tau)$ is

$$H_1(\tau) = \Phi_a^T(t_f,\tau) N_f \Phi_a(t_f,\tau) + \int_\tau^{t_f} \Phi_a^T(\sigma,\tau) N_a(\sigma) \Phi_a(\sigma,\tau) d\sigma.$$

It follows that the matrix norm of $H_1(\tau)$ can be bounded above by

$$||H_1(\tau)|| \le ||N_f|| ||\Phi_a(t_f,\tau)||^2 + \int_\tau^{t_f} ||N_a(\sigma)|| ||\Phi_a(\sigma,\tau)||^2 d\sigma$$

$$\le ||N_f|| ||\Phi_a(t_f,\tau)||^2 + \overline{N}_a \int_\tau^{t_f} ||\Phi_a(\sigma,\tau)||^2 \, d\sigma.$$

Because the system (8.16) is exponentially stable, one can always find positive constants such that

$$\lim_{t_f \to \infty} \int_\tau^{t_f} ||\Phi_a(\sigma,\tau)||^2 \, d\sigma \le M_2,$$

$$\lim_{t_f \to \infty} ||\Phi_a(t_f,\tau)|| \le \lim_{t_f \to \infty} c_1 e^{-c_2(t_f - \tau)} = 0$$

which in turn yields

$$||H_1(\tau)|| \le \overline{N}_a M_2 \triangleq \Gamma_1.$$

For $2 \leq r \leq k$, the solutions $H_r(\tau)$ can be written as follows

$$H_r(\tau) = \int_{\tau}^{t_f} \Phi_a^T(\sigma,\tau) \sum_{s=1}^{r-1} H_s(\sigma) G_a(\sigma) W_a G_a^T(\sigma) H_{r-s}(\sigma) \Phi_a(\sigma,\tau) \mathrm{d}\sigma$$

from which one can further write

$$\begin{aligned} ||H_r(\tau)|| &\leq \int_{\tau}^{t_f} \left\| \Phi_a^T(\sigma,\tau) \sum_{s=1}^{r-1} H_s(\sigma) G_a(\tau) W_a G_a^T(\sigma) H_{r-s}(\sigma) \Phi_a(\sigma,\tau) \right\| \mathrm{d}\sigma \\ &\leq \int_{\tau}^{t_f} ||\Phi_a(\sigma,\tau)||^2 \sum_{s=1}^{r-1} ||H_s(\sigma)|| ||H_{r-s}(\sigma)|| ||G_a(\sigma)||^2 ||W_a|| \mathrm{d}\sigma. \end{aligned}$$

Knowing recursively, for $1 \leq s \leq r-1$ and $2 \leq r \leq k$,

$$\begin{aligned} ||H_s(\sigma)|| &\leq \Gamma_s, \\ ||H_{r-s}(\sigma)|| &\leq \Gamma_{r-s}, \end{aligned}$$

one further obtains

$$\begin{aligned} ||H_r(\tau)|| &\leq \sum_{s=1}^{r-1} \Gamma_s \Gamma_{r-s} \overline{G}_a^2 ||W_a|| \int_{\tau}^{t_f} ||\Phi_a(\sigma,\tau)||^2 \mathrm{d}\sigma \\ &\leq \sum_{s=1}^{r-1} \Gamma_s \Gamma_{r-s} \overline{G}_a^2 ||W_a|| M_2. \end{aligned}$$

Notating $\Gamma_r \triangleq \sum_{s=1}^{r-1} \Gamma_s \Gamma_{r-s} \overline{G}_a^2 ||W_a|| M_2$, it can finally be shown that

$$||H_r(\tau)|| \leq \Gamma_r, \qquad 2 \leq r \leq k,$$

which completes the proof. □

8.3 Statements Toward the Optimal Decision Problem

Next, the use of increased insight into the roles played by performance-measure statistics on the chi-squared random performance measure (8.17) of the statistical optimal control problem with low sensitivity is the main focus of this section. To formulate in precise terms the optimization problem under consideration, it is important to note that all the performance-measure statistics are functions of time-backward evolutions and do not depend on intermediate interaction values $z(t)$.

Henceforth, the time-backward trajectories (8.29)–(8.31) are therefore considered as the new dynamical equations with the associated state variables $H_r(\tau)$ and $D_r(\tau)$, not the traditional system states $z(t)$.

For such statistical optimal control problems, the importance of a compact statement is advocated to aid mathematical manipulations. To make this more precise, one may think of the k-tuple state variables $\mathscr{H}(\cdot) \triangleq (\mathscr{H}_1(\cdot),\dots,\mathscr{H}_k(\cdot))$ and $\mathscr{D}(\cdot) \triangleq (\mathscr{D}_1(\cdot),\dots,\mathscr{D}_k(\cdot))$ whose continuously differentiable states $\mathscr{H}_r \in \mathscr{C}^1([t_0,t_f];\mathbb{R}^{4n\times 4n})$ and $\mathscr{D}_r \in \mathscr{C}^1([t_0,t_f];\mathbb{R})$ having the representations $\mathscr{H}_r(\cdot) \triangleq H_r(\cdot)$ and $\mathscr{D}_r(\cdot) \triangleq D_r(\cdot)$ with the right members satisfying the dynamics (8.29)–(8.31) are defined on $[t_0,t_f]$.

In the sequel, the bounded and Lipschitz continuous mappings are introduced as

$$\mathscr{F}_r : [t_0,t_f] \times (\mathbb{R}^{4n\times 4n})^k \mapsto \mathbb{R}^{4n\times 4n},$$
$$\mathscr{G}_r : [t_0,t_f] \times (\mathbb{R}^{4n\times 4n})^k \mapsto \mathbb{R},$$

where the rules of action are given by

$$\mathscr{F}_1(\tau,\mathscr{H}) \triangleq -F_a^T(\tau)\mathscr{H}_1(\tau) - \mathscr{H}_1(\tau)F_a(\tau) - N_a(\tau),$$
$$\mathscr{F}_r(\tau,\mathscr{H}) \triangleq -F_a^T(\tau)\mathscr{H}_r(\tau) - \mathscr{H}_r(\tau)F_a(\tau)$$
$$-\sum_{s=1}^{r-1} \frac{2r!}{s!(r-s)!}\mathscr{H}_s(\tau)G_a(\tau)W_aG_a^T(\tau)\mathscr{H}_{r-s}(\tau), \quad 2 \le r \le k,$$
$$\mathscr{G}_r(\tau,\mathscr{H}) \triangleq -\mathrm{Tr}\left\{\mathscr{H}_r(\tau)G_a(\tau)W_aG_a^T(\tau)\right\}, \quad 1 \le r \le k.$$

The product mappings that follow are necessary for a compact formulation

$$\mathscr{F}_1 \times \cdots \times \mathscr{F}_k : [t_0,t_f] \times (\mathbb{R}^{4n\times 4n})^k \mapsto (\mathbb{R}^{4n\times 4n})^k,$$
$$\mathscr{G}_1 \times \cdots \times \mathscr{G}_k : [t_0,t_f] \times (\mathbb{R}^{4n\times 4n})^k \mapsto \mathbb{R}^k,$$

whereby the corresponding notations $\mathscr{F} \triangleq \mathscr{F}_1 \times \cdots \times \mathscr{F}_k$ and $\mathscr{G} \triangleq \mathscr{G}_1 \times \cdots \times \mathscr{G}_k$ are used. Thus, the dynamic equations of motion (8.29)–(8.31) can be rewritten as

$$\frac{d}{d\tau}\mathscr{H}(\tau) = \mathscr{F}(\tau,\mathscr{H}(\tau)), \qquad \mathscr{H}(t_f) \equiv \mathscr{H}_f, \tag{8.33}$$

$$\frac{d}{d\tau}\mathscr{D}(\tau) = \mathscr{G}(\tau,\mathscr{H}(\tau)), \qquad \mathscr{D}(t_f) \equiv \mathscr{D}_f, \tag{8.34}$$

whereby $\mathscr{H}_f \triangleq (N_f,0,\dots,0)$ and $\mathscr{D}_f = (0,\dots,0)$.

Notice that the product system uniquely determines the state matrices $\mathscr{H}$ and $\mathscr{D}$ once the admissible feedback gain K being specified. Henceforth, these state variables will be considered as $\mathscr{H}(\cdot) \equiv \mathscr{H}(\cdot,K)$ and $\mathscr{D}(\cdot) \equiv \mathscr{D}(\cdot,K)$.

For the given terminal data $(t_f,\mathscr{H}_f,\mathscr{D}_f)$, the classes of admissible feedback gains are next defined.

Definition 8.3.1 (Admissible Feedback Gains). Let compact subset $\overline{K} \subset \mathbb{R}^{m\times n}$ be the set of allowable feedback gain values. For the given $k \in \mathbb{Z}^+$ and sequence $\mu = \{\mu_r \geq 0\}_{r=1}^k$ with $\mu_1 > 0$, the set of feedback gains $\mathscr{K}_{t_f,\mathscr{H}_f,\mathscr{D}_f;\mu}$ is assumed to be the class of $\mathscr{C}([t_0,t_f];\mathbb{R}^{m\times n})$ with values $K(\cdot) \in \overline{K}$ for which solutions to the dynamic equations (8.33)–(8.34) with the terminal-value conditions $\mathscr{H}(t_f) = \mathscr{H}_f$ and $\mathscr{D}(t_f) = \mathscr{D}_f$ exist on the interval of optimization $[t_0,t_f]$.

At this stage, it is quite reasonable for robust decisions and performance reliability to focus attention to higher-order performance-measure statistics as the test criteria for the requirement of performance-based reliability. Therefore, a more realistic performance index is the one considering both performance risks and values as concerned in statistical optimal control. Precisely stated, on $\mathscr{K}_{t_f,\mathscr{H}_f,\mathscr{D}_f;\mu}$, the performance index with risk-value awareness is subsequently defined as follows.

Definition 8.3.2 (Risk-Value Aware Performance Index). Fix $k \in \mathbb{Z}^+$ and the sequence of scalar coefficients $\mu = \{\mu_r \geq 0\}_{r=1}^k$ with $\mu_1 > 0$. Then, for the given z_0, the risk-value aware performance index

$$\phi_0 : \{t_0\} \times (\mathbb{R}^{4n\times 4n})^k \times \mathbb{R}^k \mapsto \mathbb{R}^+$$

pertaining to statistical optimal control of the stochastic system with low sensitivity over $[t_0,t_f]$ is defined by

$$\phi_0(t_0,\mathscr{H}(t_0),\mathscr{D}(t_0)) \triangleq \underbrace{\mu_1\kappa_1}_{\text{Value Measure}} + \underbrace{\mu_2\kappa_2 + \cdots + \mu_k\kappa_k}_{\text{Risk Measures}}$$

$$= \sum_{r=1}^{k} \mu_r[z_0^T \mathscr{H}_r(t_0) z_0 + \mathscr{D}_r(t_0)], \tag{8.35}$$

where additional design freedom by means of μ_r's utilized by the control designer with risk-averse attitudes are sufficient to meet and exceed different levels of performance-based reliability requirements, for instance, mean (i.e., the average of performance measure), variance (i.e., the dispersion of values of performance measure around its mean), skewness (i.e., the anti-symmetry of the probability density of performance measure), kurtosis (i.e., the heaviness in the probability density tails of performance measure), pertaining to closed-loop performance variations and uncertainties while the supporting solutions $\{\mathscr{H}_r(\tau)\}_{r=1}^k$ and $\{\mathscr{D}_r(\tau)\}_{r=1}^k$ evaluated at $\tau = t_0$ satisfy the dynamical equations (8.33)–(8.34).

Given that the terminal time t_f and states $(\mathscr{H}_f, \mathscr{D}_f)$, the other end condition involved the initial time t_0 and state pair $(\mathscr{H}_0, \mathscr{D}_0)$ are specified by a target set requirement.

Definition 8.3.3 (Target Set). $(t_0, \mathscr{H}_0, \mathscr{D}_0) \in \hat{\mathscr{M}}$, where the target set $\hat{\mathscr{M}}$ is closed under $[t_0,t_f] \times (\mathbb{R}^{4n\times 4n})^k \times \mathbb{R}^k$.

Now, the optimization problem is to minimize the risk-value aware performance index (8.35) over all admissible feedback gains $K = K(\cdot)$ in $\mathscr{K}_{t_f,\mathscr{H}_f,\mathscr{D}_f;\mu}$.

Definition 8.3.4 (Optimization Problem of Mayer Type). Fix $k \in \mathbb{Z}^+$ and the sequence of scalar coefficients $\mu = \{\mu_r \geq 0\}_{r=1}^k$ with $\mu_1 > 0$. Then, the optimization problem of the statistical control over $[t_0, t_f]$ is given by

$$\min_{K(\cdot) \in \mathscr{K}_{t_f, \mathscr{H}_f, \mathscr{D}_f; \mu}} \phi_0(t_0, \mathscr{H}(t_0), \mathscr{D}(t_0)), \tag{8.36}$$

subject to the dynamical equations (8.33)–(8.34) on $[t_0, t_f]$.

In the case of the optimization problem of Mayer type, it has been further suggested that an adaptation of the Mayer-form verification theorem of dynamic programming as depicted in [4] is necessary to obtain an appropriate control solution. To embed the aforementioned optimization into a larger optimal control problem, the terminal time and states $(t_f, \mathscr{H}_f, \mathscr{D}_f)$ are parameterized as $(\varepsilon, \mathscr{Y}, \mathscr{Z})$. Thus, the value function now depends on terminal condition parameterizations.

Definition 8.3.5 (Value Function). Suppose that $(\varepsilon, \mathscr{Y}, \mathscr{Z}) \in [t_0, t_f] \times (\mathbb{R}^{4n \times 4n})^k \times \mathbb{R}^k$ is given and fixed. Then, the value function $\mathscr{V}(\varepsilon, \mathscr{Y}, \mathscr{Z})$ is defined by

$$\mathscr{V}(\varepsilon, \mathscr{Y}, \mathscr{Z}) \triangleq \inf_{K(\cdot) \in \mathscr{K}_{\varepsilon, \mathscr{Y}, \mathscr{Z}; \mu}} \phi_0\left(t_0, \mathscr{H}(t_0, K), \mathscr{D}(t_0, K)\right).$$

For convention, $\mathscr{V}(\varepsilon, \mathscr{Y}, \mathscr{Z}) \triangleq \infty$ when $\mathscr{K}_{\varepsilon, \mathscr{Y}, \mathscr{Z}; \mu}$ is empty. To avoid cumbersome notation, the dependence of trajectory solutions on $K(\cdot)$ is suppressed. Next, some candidates for the value function are constructed with the help of the concept of reachable set.

Definition 8.3.6 (Reachable Set). Let the reachable set $\hat{\mathscr{Q}}$ be

$$\hat{\mathscr{Q}} \triangleq \left\{(\varepsilon, \mathscr{Y}, \mathscr{Z}) \in [t_0, t_f] \times (\mathbb{R}^{4n \times 4n})^k \times \mathbb{R}^k : \mathscr{K}_{\varepsilon, \mathscr{Y}, \mathscr{Z}; \mu} \neq \emptyset\right\}.$$

Notice that $\hat{\mathscr{Q}}$ contains a set of points $(\varepsilon, \mathscr{Y}, \mathscr{Z})$, from which it is possible to reach the target set $\mathscr{M}$ with some trajectory pairs corresponding to a continuous feedback gain. Furthermore, the value function must satisfy both a partial differential inequality and an equation at each interior point of the reachable set, at which it is differentiable.

Theorem 8.3.1 (HJB Equation for Mayer Problem). *Let $(\varepsilon, \mathscr{Y}, \mathscr{Z})$ be any interior point of the reachable set $\hat{\mathscr{Q}}$, at which the scalar-valued function $\mathscr{V}(\varepsilon, \mathscr{Y}, \mathscr{Z})$ is differentiable. Then, $\mathscr{V}(\varepsilon, \mathscr{Y}, \mathscr{Z})$ satisfies the partial differential inequality*

$$\begin{aligned} 0 \geq \frac{\partial}{\partial \varepsilon} \mathscr{V}(\varepsilon, \mathscr{Y}, \mathscr{Z}) &+ \frac{\partial}{\partial \operatorname{vec}(\mathscr{Y})} \mathscr{V}(\varepsilon, \mathscr{Y}, \mathscr{Z}) \operatorname{vec}(\mathscr{F}(\varepsilon, \mathscr{Y}, K)) \\ &+ \frac{\partial}{\partial \operatorname{vec}(\mathscr{Z})} \mathscr{V}(\varepsilon, \mathscr{Y}, \mathscr{Z}) \operatorname{vec}(\mathscr{G}(\varepsilon, \mathscr{Y})) \end{aligned} \tag{8.37}$$

for all $K \in \overline{\mathscr{K}}$ *and* $\text{vec}(\cdot)$ *the vectorizing operator of enclosed entities.*

If there is an optimal feedback gain K^* *in* $\mathscr{K}_{\varepsilon,\mathscr{Y},\mathscr{Z};\mu}$, *then the partial differential equation of dynamic programming*

$$0 = \min_{K\in\overline{K}} \Big\{ \frac{\partial}{\partial \text{vec}(\mathscr{Y})} \mathscr{V}(\varepsilon,\mathscr{Y},\mathscr{Z}) \text{vec}(\mathscr{F}(\varepsilon,\mathscr{Y},K)) + \frac{\partial}{\partial \text{vec}(\mathscr{Z})} \mathscr{V}(\varepsilon,\mathscr{Y},\mathscr{Z}) \text{vec}(\mathscr{G}(\varepsilon,\mathscr{Y})) + \frac{\partial}{\partial \varepsilon} \mathscr{V}(\varepsilon,\mathscr{Y},\mathscr{Z}) \Big\} \quad (8.38)$$

is satisfied. The minimum in Eq. (8.38) is achieved by the optimal feedback gain $K^*(\varepsilon)$ *at* ε.

Proof. Interested readers are referred to the mathematical details in [5]. □

The Mayer-form verification theorem in statistical optimal control notations is stated.

Theorem 8.3.2 (Verification Theorem). *Fix* $k \in \mathbb{Z}^+$ *and let* $\mathscr{W}(\varepsilon,\mathscr{Y},\mathscr{Z})$ *be a continuously differentiable solution of the HJB equation (8.38), which satisfies the boundary condition*

$$\mathscr{W}(t_0,\mathscr{H}(t_0),\mathscr{D}(t_0)) = \phi_0(t_0,\mathscr{H}(t_0),\mathscr{D}(t_0)), \quad \text{for some } (t_0,\mathscr{H}(t_0),\mathscr{D}(t_0)) \in \hat{\mathscr{M}}.$$

Let $(t_f,\mathscr{H}_f,\mathscr{D}_f)$ *be a point of* $\hat{\mathscr{Q}}$, *let* K *be a feedback gain in* $\mathscr{K}_{t_f,\mathscr{H}_f,\mathscr{D}_f;\mu}$ *and let* $\mathscr{H}(\cdot)$ *and* $\mathscr{D}(\cdot)$ *be the corresponding solutions of Eqs. (8.33)–(8.34). Then,* $\mathscr{W}(\tau,\mathscr{H}(\tau),\mathscr{D}(\tau))$ *is a non-increasing function of* τ. *If* K^* *is a feedback gain in* $\mathscr{K}_{t_f,\mathscr{H}_f,\mathscr{D}_f;\mu}$ *defined on* $[t_0,t_f]$ *with the corresponding solutions* $\mathscr{H}^*(\cdot)$ *and* $\mathscr{D}^*(\cdot)$ *of the preceding equations such that, for* $\tau \in [t_0,t_f]$,

$$0 = \frac{\partial}{\partial \varepsilon} \mathscr{W}(\tau,\mathscr{H}^*(\tau),\mathscr{D}^*(\tau)) + \frac{\partial}{\partial \text{vec}(\mathscr{Y})} \mathscr{W}(\tau,\mathscr{H}^*(\tau),\mathscr{D}^*(\tau)) \text{vec}(\mathscr{F}(\tau,\mathscr{H}^*(\tau),K^*(\tau))) + \frac{\partial}{\partial \text{vec}(\mathscr{Z})} \mathscr{W}(\tau,\mathscr{H}^*(\tau),\mathscr{D}^*(\tau)) \text{vec}(\mathscr{G}(\tau,\mathscr{H}^*(\tau))) \quad (8.39)$$

then, K^* *is an optimal feedback gain, and* $\mathscr{W}(\varepsilon,\mathscr{Y},\mathscr{Z}) = \mathscr{V}(\varepsilon,\mathscr{Y},\mathscr{Z})$, *where* $\mathscr{V}(\varepsilon,\mathscr{Y},\mathscr{Z})$ *is the value function.*

Proof. Starting from the detailed analysis in [5], the proof is readily established. □

8.4 Risk-Averse Control Solution in the Closed-Loop System

Recall that the optimization problem being considered herein is in Mayer form, which can be solved by an adaptation of the Mayer-form verification theorem. Thus, the terminal time and states $(t_f, \mathscr{H}_f, \mathscr{D}_f)$ are parameterized as $(\varepsilon, \mathscr{Y}, \mathscr{Z})$ for a family of optimization problems. For instance, the states (8.33)–(8.34) defined on the interval $[t_0, \varepsilon]$ now have terminal values denoted by $\mathscr{H}(\varepsilon) \equiv \mathscr{Y}$ and $\mathscr{D}(\varepsilon) \equiv \mathscr{Z}$, where $\varepsilon \in [t_0, t_f]$. Furthermore, with $k \in \mathbb{Z}^+$ and $(\varepsilon, \mathscr{Y}, \mathscr{Z})$ in $\hat{\mathscr{Q}}$, the following real-value candidate:

$$\mathscr{W}(\varepsilon, \mathscr{Y}, \mathscr{Z}) = z_0^T \sum_{r=1}^{k} \mu_r(\mathscr{Y}_r + \mathscr{E}_r(\varepsilon)) z_0 + \sum_{r=1}^{k} \mu_r(\mathscr{Z}_r + \mathscr{T}_r(\varepsilon)) \tag{8.40}$$

for the value function is therefore differentiable. The time derivative of $\mathscr{W}(\varepsilon, \mathscr{Y}, \mathscr{Z})$ can also be shown of the form

$$\begin{aligned}\frac{d}{d\varepsilon}\mathscr{W}(\varepsilon, \mathscr{Y}, \mathscr{Z}) &= z_0^T \sum_{r=1}^{k} \mu_r \left(\mathscr{F}_r(\varepsilon, \mathscr{Y}, K) + \frac{d}{d\varepsilon}\mathscr{E}_r(\varepsilon) \right) z_0 \\ &\quad + \sum_{r=1}^{k} \mu_r \left(\mathscr{G}_r(\varepsilon, \mathscr{Y}) + \frac{d}{d\varepsilon}\mathscr{T}_r(\varepsilon) \right),\end{aligned}$$

whereby the time-parametric functions $\mathscr{E}_r \in \mathscr{C}^1([t_0, t_f]; \mathbb{R}^{4n \times 4n})$ and $\mathscr{T}_r \in \mathscr{C}^1([t_0, t_f]; \mathbb{R})$ are to be determined.

At the boundary condition, it requires that

$$\mathscr{W}(t_0, \mathscr{H}(t_0), \mathscr{D}(t_0)) = \phi_0(t_0, \mathscr{H}(t_0), \mathscr{D}(t_0))$$

which in turn leads to

$$\begin{aligned}& z_0^T \sum_{r=1}^{k} \mu_r(\mathscr{H}_r(t_0) + \mathscr{E}_r(t_0)) z_0 + \sum_{r=1}^{k} \mu_r(\mathscr{D}_r(t_0) + \mathscr{T}_r(t_0)) \\ &= z_0^T \sum_{r=1}^{k} \mu_r \mathscr{H}_r(t_0) z_0 + \sum_{r=1}^{k} \mu_r \mathscr{D}_r(t_0).\end{aligned} \tag{8.41}$$

By matching the boundary condition (8.41), it yields that the time parameter functions $\mathscr{E}_r(t_0) = 0$ and $\mathscr{T}_r(t_0) = 0$ for $1 \le r \le k$. Next, it is necessary to verify that this candidate value function satisfies (8.39) along the corresponding trajectories produced by the feedback gain K resulting from the minimization in Eq. (8.38). Or equivalently, one obtains

$$0 = \min_{K \in \overline{K}} \left\{ z_0^T \left[\sum_{r=1}^{k} \mu_r \mathscr{F}_r(\varepsilon, \mathscr{Y}, K) + \sum_{r=1}^{k} \mu_r \frac{d}{d\varepsilon} \mathscr{E}_r(\varepsilon) \right] z_0 + \sum_{r=1}^{k} \mu_r \mathscr{G}_r(\varepsilon, \mathscr{Y}) + \sum_{r=1}^{k} \mu_r \frac{d}{d\varepsilon} \mathscr{T}_r(\varepsilon) \right\}. \tag{8.42}$$

Now the aggregate matrix coefficients $F_a(t)$ and $N_a(t)$ for $t \in [t_0, t_f]$ of the composite dynamics (8.16) with trajectory sensitivity consideration are next partitioned to conform with the n-dimensional structure of Eq. (8.1) by means of

$$I_0^T \triangleq \left[I\,0\,0\,0 \right], \quad I_1^T \triangleq \left[0\,I\,0\,0 \right], \quad I_2^T \triangleq \left[0\,0\,I\,0 \right], \quad I_3^T \triangleq \left[0\,0\,0\,I \right],$$

where I is an $n \times n$ identity matrix and

$$\begin{aligned} F_a(t) &= I_0(A(t,\zeta) + B(t,\zeta)K(t))I_0^T + I_0 L(t)C(t)I_1^T + I_1(A(t,\zeta) - L(t)C(t))I_1^T \\ &\quad + I_2 L_\zeta(t)C(t)I_1^T + I_2(A_\zeta(t,\zeta_0) + B_\zeta(t,\zeta_0)K(t))I_0^T + I_2 L(t)C(t)I_3^T \\ &\quad + I_2(A(t,\zeta_0) + B(t,\zeta_0)K(t))I_2^T + I_3(A_\zeta(t,\zeta_0) - L_\zeta(t)C(t))I_1^T \\ &\quad + I_3(A(t,\zeta) - L(t)C(t))I_3^T , \end{aligned} \tag{8.43}$$

$$\begin{aligned} N_a(t) &= I_0(Q(t) + K^T(t)R(t)K(t))I_0^T + I_0 Q(t)I_1^T + I_1 Q(t)I_0^T + I_1 Q(t)I_1^T \\ &\quad + I_2 S(t)I_2^T + I_2 S(t)I_3^T + I_3 S(t)I_2^T + I_3 S(t)I_3^T . \end{aligned} \tag{8.44}$$

Therefore, the derivative of the expression in Eq. (8.42) with respect to the admissible feedback gain K yields the necessary conditions for an extremum of Eq. (8.38) on $[t_0, t_f]$,

$$K = -R^{-1}(\varepsilon)[B^T(\varepsilon,\zeta)I_0^T + B_\zeta^T(\varepsilon,\zeta_0)I_2^T] \cdot \sum_{s=1}^{k} \hat{\mu}_s \mathscr{Y}_s I_0((I_0^T I_0)^{-1})^T , \tag{8.45}$$

whereby the normalized degrees of freedom $\hat{\mu}_s = \frac{\mu_s}{\mu_1}$ with $\mu_1 > 0$.

With the feedback gain (8.45) replaced in the expression of the bracket (8.42) and having $\{\mathscr{Y}_s\}_{s=1}^k$ evaluated on the solution trajectories (8.33)–(8.34), the time dependent functions $\mathscr{E}_r(\varepsilon)$ and $\mathscr{T}_r(\varepsilon)$ are therefore chosen such that the sufficient condition (8.39) in the verification theorem is satisfied in the presence of the arbitrary value of z_0, e.g.,

$$\frac{d}{d\varepsilon} \mathscr{E}_1(\varepsilon) = F_a^T(\varepsilon)\mathscr{H}_1(\varepsilon) + \mathscr{H}_1(\varepsilon)F_a(\varepsilon) + N_a(\varepsilon),$$

$$\frac{d}{d\varepsilon}\mathscr{E}_r(\varepsilon) = F_a^T(\varepsilon)\mathscr{H}_r(\varepsilon) + \mathscr{H}_r(\varepsilon)F_a(\varepsilon) + \sum_{s=1}^{r-1}\frac{2r!}{s!(r-s)!}\mathscr{H}_s(\varepsilon)G_a(\varepsilon)W_aG_a^T(\varepsilon)\mathscr{H}_{r-s}(\varepsilon), \quad 2 \le r \le k,$$

$$\frac{d}{d\varepsilon}\mathscr{T}_r(\varepsilon) = \mathrm{Tr}\{\mathscr{H}_r(\varepsilon)G_a(\varepsilon)W_aG_a^T(\varepsilon)\}, \quad 1 \le r \le k,$$

whereby the initial-value conditions $\mathscr{E}_r(t_0) = 0$ and $\mathscr{T}_r(t_0) = 0$ for $1 \le r \le k$. Therefore, the sufficient condition (8.39) of the verification theorem is satisfied so that the extremizing feedback gain (8.45) becomes optimal.

In short, it has been shown that the optimal stochastic linear system with risk-averse output feedback and low sensitivity is summarized as follows.

Theorem 8.4.1 (Risk-Averse Output Feedback Control with Low Sensitivity). *Let (A,B) and (A,C) be uniformly stablizable and detectable. Consider $u(t) = K(t)\hat{x}(t)$ for all $t \in [t_0,t_f]$, where the state estimates $\hat{x}(t)$ is governed by the dynamical observer (8.9). Then, the low sensitivity control strategy with risk aversion is supported by the output-feedback gain $K^*(\cdot)$*

$$K^*(t) = -R^{-1}(t)[B^T(t,\zeta)I_0^T + B_\zeta^T(t,\zeta_0)I_2^T]\sum_{s=1}^{k}\hat{\mu}_s\mathscr{H}_s^*(t)I_0((I_0^TI_0)^{-1})^T, \quad (8.46)$$

where $t \triangleq t_0 + t_f - \tau$ and the normalized parametric design of freedom $\hat{\mu}_s \triangleq \frac{\mu_r}{\mu_1}$. The optimal state solutions $\{\mathscr{H}_r^*(\cdot)\}_{r=1}^{k}$ are supporting the mathematical statistics for performance robustness and governed by the backward-in-time matrix-valued differential equations with the terminal-value conditions $\mathscr{H}_1^*(t_f) = N_f$ and $\mathscr{H}_r^*(t_f) = 0$ for $2 \le r \le k$

$$\frac{d}{d\tau}\mathscr{H}_1^*(\tau) = -(F_a^*)^T(\tau)\mathscr{H}_1^*(\tau) - \mathscr{H}_1^*(\tau)F_a^*(\tau) - N_a^*(\tau), \quad (8.47)$$

$$\frac{d}{d\tau}\mathscr{H}_r^*(\tau) = -(F_a^*)^T(\tau)\mathscr{H}_r^*(\tau) - \mathscr{H}_r^*(\tau)F_a^*(\tau) - \sum_{s=1}^{r-1}\frac{2r!}{s!(r-s)!}\mathscr{H}_s^*(\tau)G_a(\tau)W_aG_a^T(\tau)\mathscr{H}_{r-s}^*(\tau), \quad 2 \le r \le k \quad (8.48)$$

while the Kalman-filtering gain $L(t)$ and its partial $L_\zeta(t)$ with respect to the constant parameter ζ are governed by the estimate-error covariance equations (8.11) and (8.15) on $[t_0,t_f]$.

8.5 Chapter Summary

An optimal risk-averse control strategy with low sensitivity and subject to noisy feedback measurements is investigated. The conditional mean estimates of the current states are generated by the use of a Kalman filter. The risk-averse gain of output feedback is obtained by solving the set of backward-in-time matrix differential equations that are used to compute the mathematical statistics associated with performance uncertainty, e.g., mean, variance, and skewness. In simple terms, the present results provide an enhanced theory for stochastic control with performance-information analysis whereby responsive risk-value aware performance indices are incorporated for emerging applications. More precisely, major contributions include: (1) complete characterization and management of value of information pertaining to mathematical statistics associated with quadratic utilities; (2) a global analysis approach to an unified framework of measuring risk judgments and modeling choices and decisions; and (3) a dynamic programming adapted for determining solutions. Finally, the results here are certainly viewed as the generalization of the previous chapter.

References

1. D'Angelo, H., Moe, M.L., Hendricks, T.C.: Trajectory sensitivity of an optimal control systems. In: Proceedings of the 4th Allerton Conference on Circuit and Systems Theory, pp. 489–498 (1966)
2. Kahne, S.J.: Low-sensitivity design of optimal linear control systems. IEEE Trans. Aero. Electron. Syst. **4**(3), 374–379 (1968)
3. Pollatsek, A., Tversky, A.: Theory of risk. J. Math. Psychol. **7**, 540–553 (1970)
4. Fleming, W.H., Rishel, R.W.: Deterministic and Stochastic Optimal Control. Springer, New York (1975)
5. Pham, K.D.: Performance-reliability aided decision making in multiperson quadratic decision games against jamming and estimation confrontations. In: Giannessi, F. (ed.) J. Optim. Theor. Appl. **149**(3), 559–629 (2011)
6. Pham, K.D.: New risk-averse control paradigm for stochastic two-time-scale systems and performance robustness. In: Miele, A. (ed.) J. Optim. Theor. Appl. **146**(2), 511–537 (2010)

Chapter 9
Epilogue

Abstract On a closing note, the purpose here is to shed some light on two striking perspectives, e.g., (1) the acquisition and utilization of insights, regarding whether the control designers are risk averse or risk prone and thus restrictions on the utility functions implied by these attitudes, to adapt feedback control strategies to meet austere environments and (2) best courses of action to ensure performance robustness and reliability, provided that the control designers be subscribed to certain attitudes that all the chapters in this monograph share and contribute defining. In addition, this research initiative will serve as the first brick in the larger, interwoven wall of the development and practice of policies, standards, and procedures that would enable greater efficiency worthiness designed into future controlled systems as further improvements in reliability occur and hence underline autonomous system acceptance into system analysis and control synthesis communities.

9.1 The Aspect of Performance-Information Analysis

Given the lack of analysis of performance risk and stochastic preferences, the research chapters contained in this monograph illustrated well the performance-information analysis and interpretations of what they mean for performance riskiness from the standpoint of higher-order characteristics pertaining to performance sampling distributions. And, for the reason of measure of effectiveness, much of the discussion from the monograph has concerned the case involving the behavior of performance measure. In such situations of stabilization and regulation where the control designers have risk-averse attitudes toward performance uncertainty, they are thus interested in a utility function whose property of the utility of the expected performance of any closed-loop implementation be greater than the expected utility of that closed-loop execution. Hence, it is quite reasonable to consider the utility to have the concave shape and functional form, e.g., the logarithm of the cost characteristic function.

K.D. Pham, *Linear-Quadratic Controls in Risk-Averse Decision Making*, SpringerBriefs in Optimization, DOI 10.1007/978-1-4614-5079-5_9,

In this monograph, the framing chapters suggested the question of how to characterize and influence closed-loop performance information to be answered by modeling and management of cumulants (also known as semi-invariants). Precisely stated, these cumulants can be generated by the use of the Maclaurin series expansion of the concave shape utility function. The technical results provided an effective and accurate capability for forecasting all the higher-order characteristics associated with performance uncertainty. Therefore, via higher-order cost cumulants and adaptive decision making, it is anticipated that future performance variations will lose the element of surprise due to the inherent property of self-enforcing and risk-averse control solutions, that are highly capable of reshaping the cumulative probability distribution of closed-loop system performance.

9.2 The Aspect of Risk-Averse Decision Making

Paradoxically, in much of the optimization, estimation and control work relevant to onboard automation and autonomy with particular emphases on "correct-by-design" performance reliability, robust control, and decision making under performance uncertainty and risk is seldomly addressed beyond the widely used measure of statistical average or mean to summarize the underlying performance variations. The emerging problems examined are then simply the new model concepts for performance benefits and risks, which essentially consist of trade-offs between performance values and risks for reliable decisions. It therefore underlies a risk-aversion notion in performance assessment and feedback decision design under uncertainty.

As repeatedly emphasized throughout this monograph, the controller designers are no longer content with the averaged performance and thus decide to mitigate performance riskiness by the use of a new performance index incorporated with performance values and risks to ensure how much of the inherent or design-in reliability actually ends up in the developmental and operational phases. Therefore, it is expected that risk-averse feedback control strategies developed here are able to foresee and hedge the changes in system performance outcomes that advance hence the uncertain environment's mixed random realizations.

9.3 What Will Be?

In conclusion, the statistical optimal control theory of stochastic linear-quadratic systems contributes to the applications of performance-information analysis and measures of risk prone and risk aversion in feedback control design. As anticipated, these foundational results herein will augment self-correcting and risk-averse attributes to autonomous systems in situations whereby human response time is insufficient and severe faults happen. It essentially provides control engineers with the technical background and design algorithms that facilitate an efficient solution of stochastic control problems with performance reliability requirements. The chapters in this monograph offer hope for a growing science and technology. If the control practitioners and theorists are serious about exploring the quantifications of closed-loop system performance of what might be, developing a more elaborate science of the normative, and if they want to know more about what lies beyond what is and explore what might be, the future is bright.

Index

A
a-priori scheduling signals, 31
adaptive behavior, 8
adaptive control decisions, 32
admissible controls, 50, 90, 109, 126, 127
admissible feedback gains, 9, 51, 69, 115, 128, 137
admissible feedforward gains, 51, 90
admissible initial condition, 8, 33, 69
admissible set, 77, 97, 126

B
backward-in-time matrix-valued differential equations, 28, 46, 53, 70, 113, 122, 143
backward-in-time scalar-valued differential equations, 71
backward-in-time vector-valued differential equations, 28, 46, 71, 91

C
chi-squared, 9, 45, 50, 73, 110, 136
closed-loop performance, 113, 130
conditional probability density, 127
continuous-time coefficients, 8, 68, 108
controlled stochastic differential equations, 68, 90, 109, 129
cost cumulants, 7, 34, 53, 68, 71, 88, 110, 130
cumulant-generating function, 9, 34, 70, 91, 110, 130
cumulative probability distribution, 113, 130

D
decision process, 67, 89
design of freedom, 19, 40, 56, 77, 97, 116, 122, 138
desired trajectory, 7, 31, 64
deterministic systems, 31
dynamic programming, 20, 32, 50, 77, 100, 118, 126

E
estimate-error covariance, 128
estimate-state mean, 128

F
finite horizon, 7, 32, 49, 67, 87, 115, 126
forward-in-time matrix-valued differential equations, 122

G
goal seeking, 8

H
Hamilton-Jacobi-Bellman equation, 20, 41, 58, 78, 101, 140
Hamiltonian functional, 121
Hessian matrix, 121
Hilbert space, 8, 33, 50

I
incomplete information, 126
information process, 50, 67, 88
initial-value condition, 8, 63, 84, 104, 143
integral-quadratic-form cost, 7, 51, 69, 89, 108, 127

K
Kalman-like estimator, 125
Kronecker matrix product operator, 121

K.D. Pham, *Linear-Quadratic Controls in Risk-Averse Decision Making*, SpringerBriefs in Optimization, DOI 10.1007/978-1-4614-5079-5,

L
Linear-Quadratic-Gausian, 7
local extremum, 120
long-run average performance, 65, 68
low sensitivity, 107, 125

M
Maclaurin series, 71, 91, 112, 130
matrix minimum principle, 120
matrix variation of constants formula, 121
Mayer problem, 16, 32, 58, 78, 97
Mayer-form verification theorem, 21, 42, 59, 80, 139
model-following, 49, 68
moment-generating function, 9, 34, 70, 111, 130

O
output feedback, 125
overtaking tracking, 31

P
parameter sensitivity, 107
performance information, 31
performance reliability, 56, 67, 88, 126
performance risk aversion, 8, 38, 107, 130
performance risks, 18, 50, 85, 87, 138
performance robustness, 32, 49, 87, 108, 122, 125, 126
performance uncertainty, 12, 32, 49, 68, 88, 107, 112, 125
performance values, 17, 76, 88, 116, 132
performance-measure statistics, 8, 31, 32, 38, 52, 71, 91, 112, 132
post controller-design analysis, 7
post-design performance assessment, 31

R
random cost, 12, 108
reachable set, 20, 41, 57, 78, 99, 139
reference control, 33
reference trajectory, 46
Risk Sensitive Control, 8
risk-averse control, 7, 34, 50, 67, 87, 113, 144
risk-averse decision making, 31, 116
risk-averse output feedback, 143
risk-value aware performance index, 9, 33, 76, 116, 138

S
sample-path realizations, 32
second-order statistic, 46, 131
square-integrable process, 8
stationary Wiener process, 8, 49, 68, 108, 127
statistical optimal control, 7, 31, 49, 107, 128
stochastic disturbances, 31, 32, 68
stochastic effects, 127
stochastic elements, 49
stochastic linear systems, 107, 125
stochastic preferences, 96, 130
stochastic servo systems, 56, 67, 87
stochastic tracking systems, 8, 31

T
target probability density function, 47
terminal penalty, 8, 127
terminal-value conditions, 10, 35, 52, 72, 92, 112, 122, 130
three-degrees-of-freedom, 68
trajectory sensitivity, 107, 125
two-degrees-of-freedom, 67
two-point boundary conditions, 120

V
value function, 20, 32, 57, 77, 97, 139